C0-ALY-696

LMDS

Other McGraw-Hill Telecommunication Books of Interest

LMDS

Clint Smith, P.E.

McGraw-Hill
New York   San Francisco   Washington, D.C.
Auckland   Bogotá   Caracas   Lisbon   London
Madrid   Mexico City   Milan   Montreal   New Delhi
San Juan   Singapore   Sydney   Tokyo   Toronto

McGraw-Hill

A Division of The McGraw-Hill Companies

Copyright © 2000 by The McGraw-Hill Companies, Inc. All rights reserved. Printed in the United States of America. Except as permitted under the United States Copyright Act of 1976, no part of this publication may be reproduced or distributed in any form or by any means, or stored in a data base or retrieval system, without the prior written permission of the publisher.

1 2 3 4 5 6 7 8 9 0   DOC/DOC   9 0 9 8 7 6 5 4 3 2 1 0 9

ISBN 0-07-136254-1

The sponsoring editor for this book was Steve Chapman, the editing supervisor was Sally Glover, and the production supervisor was Pamela Pelton. It was set in Century Schoolbook by D&G Limited, LLC.

Printed and bound by R.R. Donnelley & Sons Co.

McGraw-Hill books are available at special quantity discounts to use as premiums and sales promotions, or for use in corporate training programs. For more information, please write to Director of Special Sales, McGraw-Hill, Two Penn Plaza, New York, NY 10121-2298. Or contact your local bookstore.

Information contained in this work has been obtained by The McGraw-Hill Companies, Inc. ("McGraw-Hill") from sources believed to be reliable. However, neither McGraw-Hill nor its authors guarantees the accuracy or completeness of any information published herein and neither McGraw-Hill nor its authors shall be responsible for any errors, omissions, or damages arising out of use of this information. This work is published with the understanding that McGraw-Hill and its authors are supplying information but are not attempting to render engineering or other professional services. If such services are required, the assistance of an appropriate professional should be sought.

This book is printed on recycled, acid-free paper containing a minimum of 50 percent recycled de-inked fiber.

621.3845
S644l

This book is again dedicated to Mary, Sam, and Rose for their unwavering support for all my activities.

CONTENTS

Contents

Contents

Contents

PREFACE

LMDS or FWPMP, depending on which continent this is referenced from, is fostering a revolution in communications. Cellular was fostered in the era of mobility, and LMDS is an enabler for broadband services both in developed and undeveloped countries. At present there is a race on in telecommunications in the delivery of broadband services. There are many technology platforms which exist now or will exist shortly which are all meant to unravel the broadband genie. However, each platform has its advantages and disadvantages either from physical media, bandwidth, or capital requirements.

LMDS has the distinct capability of being a key enabler for broadband services. There is a plethora of LMDS platforms and each platform has advantages and disadvantages, depending on the application or service being deliverd. LMDS as itself can operate over a wide selection of frequency bands, has multiple transport protocols which are used, and can handle a vast array of services. The deployment of a LMDS system is more precision targeted then the blanket coverage pursued by Cellular and PCS operators with mobility.

With all the activity in the LMDS/FWPMP arena, there is scarce documentation which attempts to define how to design a LMDS system and then operate it. As with any new technology or technology platform, documentation and guidelines are the last item on the list of items to put together as you roll out a system. This book is meant to provide the technical guidance for designing and deploying a LMDS/FWPMP utilizing previous experiences.

This book will assist middle and upper management, engineering, and new entrants into the LMDS and FWPMP field. This book covers a vast array of material and is meant to assist in the technical community with making informed decisions involving design aspects as well as many issues related to deployment.

The book is organized so that it takes you from technology decisions to which platform to choose and services to offer. As part of planning, the interaction between the marketing department and the technical group is covered through the various planning aspects, and then looks at what information needs to be delivered to the technical community.

The RF design and Network design are also addressed. Both designs are related, and their relationship is shown with what topics need to be addressed. The Host Terminal or Edge Terminal which the customer connects to is also discussed, with focus on multiple dwelling units for the purpose of spreading the capital and operating expenses. Implementation of all the platforms is also covered, along with the technical organization structure with the required headcount drivers. Various reports are also included for possible use, along with spectrum allocation tables and a technical glossary for some terms which may not be included in this book.

Overall, I hope that you find this book useful in that you can take many of the items covered and implement them into your LMDS design.

—CLINT SMITH, P.E.

Introduction to LMDS

Introduction

The push for elimination of the wired local loop has fostered numerous radio technologies that operate over a vast range of spectrum from 400MHz to 40GHz. The initial concepts of wireless access involved delivering voice services, whether they were analog or digitized voice, utilizing a host of modulation techniques. The primary focus was on the deployment of radio base stations and then the development of adjunct services from which customer retention and enhanced revenues could be exploited. As the Internet and other bandwidth-hungry services and products have become more prolific in society, however, data followed by voice services are what is envisioned.

With the proliferation of IP and its permutation into all aspects of life, both business and personal, the need to support additional services has become the driving force for all wired and wireless technologies. It is hard for any day to go by without some reference to the Internet or IP technology, with proclamations on how it is revolutionizing society. Society is being revolutionized through access to information for the purposes of making business and purchasing decisions. The information that is available and is becoming available is so vast that traditional concepts of time and location are no longer applicable. But the revolution is not immediate—it is expected to take several years to fully unfold due to the current bandwidth bottleneck that exists at the last mile or kilometer of any system.

Fortunately (or unfortunately), there are numerous technology platforms being deployed or considered, which will provide the needed bandwidth to facilitate the information revolution that is taking place. The difficulty in the technology platform race is to not only determine which transport medium meets the current demand but also any future demand.

One of the most promising technology types for providing broadband access is *Local Multipoint Distribution System* (LMDS). LMDS is a term that can apply to numerous types of technology platforms. LMDS offers the capability to deliver not only voice services, but also high-speed data—thereby providing a complete package of functions and services.

From the customers' aspect, the system should appear and operate with equal or better reliability than that of the local PSTN/PTT. They expect the system to not only be reliable, but also offer more features and functions while providing cost savings that are real and sustained. Therefore, the key players in any broadband system are the services that can be successfully and economically offered. There are numerous LMDS technology platforms to choose from that can meet the challenge—based on the market, spectrum available, and services offered.

Over the next few chapters, I will attempt to distill numerous aspects pertaining to the design, deployment, and operation of an LMDS system. As you will come to see, there are a vast number of LMDS systems from which to choose. As most issues unfold, however, the more unique types of systems there are to choose from, the more they have in common. It is through understanding the commonality and the differences between different LMDS and wireless systems that proper decisions can be made regarding a new or existing system.

Communication History

Data communication began in 1844, when Samuel Morse invented and pioneered the telegraph that used Morse Code as its method for delivering communication over vast distances by interweaving dits and dashes. The method of coding was so good that it is still used extensively throughout the world. The use of data communication moved from a wired to a wireless service, thanks to the efforts of Marconi, who is credited with inventing radio.

A very brief summary of some of the major milestones of the telecommunication industry is shown as follows for quick reference. There are numerous other milestones that are of extreme importance for communications, but the list represents some of the major advances.

1844—Samuel Morse invents the telegraph.

1876—Alexander Bell invents the telephone.

1901—Marconi sends Morse Code by using a radio.

1931—First U.S. television transmission takes place.

1946—AT&T offers mobile phone service.

1953—First Microwave network is installed.

1956—Transatlantic cable is constructed.

1977—Bell Labs transmits TV signals on optical fibers.

1983—Cellular Communication fosters a communications revolution.

Wireless, as it goes, has not been in existence long. The first systems were both two-way as well as broadcast. The radio communication initially focused primarily on voice communication, but after the advent of television, it also delivered broadband video with some data for instructing the television on how to display the picture. Microwave communications delivered high-speed data, in which speeds of 155Mbps are now common.

The wireline offerings have also taken a major leap from offering voice. Data applications initially involved being able to use a modem that operated at 300 baud, they then progressed to 1200 baud, and they now have speeds that exceed 1Gbps—with new speeds being reached every year as the need for more throughput increases.

Part of the need for more throughput culminated in the invention of the fax machine and the proliferation of the computer, and its numerous technological advances created the need for increased data throughput. The Internet has fostered the need for increased bandwidth, so a host of services still being dreamed of can be delivered to subscribers.

Both voice and data are merging into similar transport platforms that require not only bandwidth, but also design and management skills. The communication future at this time appears to be heading toward IP as the primary edge or end user protocol, with other supporting transport protocols such as ATM being used to deliver the information between similar and dissimilar networks.

Cellular

Cellular communication is one of the most prolific voice communication platforms deployed within the last two decades. Cellular has always been able to transport data, and many advancements in different modulation formats foster the deliver of narrow band data services. Broadband data services, however, are not achievable with cellular systems because of bandwidth limitations. Typical data rates experienced by cellular applications are 9.6kbps.

Over all, cellular communication is the form of wireless communication that enables the following key concepts to be employed:

- Frequency reuse
- Mobility of the subscriber
- Handoffs

The cellular concept is employed in many different forms. Typically, when referencing cellular communication, it is first applied to either the AMPS or TACS technology. AMPS operates in the 800MHz band (821–849MHz) base station receive and (869–894MHz) for the base station transmit. For TACS, the frequency range is 890MHz–915Mhz base receive and 935MHz–960MHz base station transmit.

Many other technologies also fall within the guise of cellular communication, and they involve the PCS bands (both domestic U.S. and the international bands). In addition, the same concept is applied to several technology platforms that are currently used in the SMR band (IS-136 and IDEN). Cellular communication is really referenced to both the AMPS and TACS bands, but it is sometimes interchanged with the PCS and SMR bands because of the similarities.

The concept of cellular radio was initially developed by AT&T at its Bell Laboratories to provide additional radio capacity for a geographic customer service area. The initial mobile systems from which cellular evolved were called *Mobile Telephone Systems* (MTS). Later, improvements to these systems occurred and the systems were referred to as *Improved Mobile Telephone Systems* (IMTS). One of the main problems with these systems is that a mobile call could not be transferred from one radio station to another without loss of communication. This problem was resolved by implementing the concept of reusing the allocated frequencies of the system. Reusing the frequencies in cellular systems enables a market to offer higher radio traffic capacity. The increased radio traffic allows more users in a geographic service area than with the MTS or IMTS systems.

Cellular radio was a logical progression in the quest to provide additional radio capacity for a geographic area. The cellular system as it is known today has its primary roots in the MTS and IMTS. Both MTS and IMTS are similar to cellular, except that no hand-off takes place with these networks.

Cellular systems operate on the principle of frequency reuse. Frequency reuse in a cellular market gives a cellular operator the capability to offer higher radio traffic capacity. The higher radio traffic capacity enables many more users in a geographic area to utilize radio communication than are available with a MTS or IMTS system.

The cellular systems in the United States are broken into *Metropolitan Statistical Areas* (MSAs) and *Rural Statistical Areas* (RSAs). Each MSA and RSA has two different cellular operators that offer service. The two cellular operators are referred to as A-band and B-band systems. The A-band system is the non-wireline system, and the B-band is the wireline system for the MSA or RSA.

A brief system configuration for a cellular system is shown in Figure 1-1.

There are numerous types of cellular systems that are used both in the United States and elsewhere. The following is a brief listing of some of the more common cellular systems. All of the systems are similar in network layout because they have base stations connected to an MSC, which in turn connects to the PSTN or PTT.

Figure 1-1
Generic cellular
system

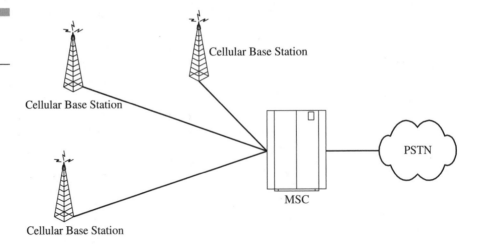

1. **Advanced Mobile Phone System** (*AMPS*) *standard*   AMPS is the cellular standard that was developed for use in North America. This type of system operates in the 800Mhz frequency band. AMPS systems have also been deployed in South America, Asia, and Russia.

2. **Code Division Multiple Access** (*CDMA*)   CDMA is an alternative digital cellular standard developed in the United States. CDMA utilizes the IS−95 standard and is implemented as the next generation for cellular systems. The CDMA system coexists with the current analog system.

3. **Digital AMPS System** (*D-AMPS*)   Also called *North American Digital Cellular* (NADC), D-AMPS is the digital standard for cellular systems developed for use in the United States. Rather than develop a completely new standard, the AMPS standard was developed into the D-AMPS digital standard. This was done to quickly provide a means to expand the existing analog systems that were growing at a rapid pace. NADC is designed to coexist with current cellular systems and relies on both the IS−54 and IS−136 standards.

4. **Global System for Mobile Communications** (*GSM*)   GSM is the European standard for digital cellular systems operating in the 900MHz band. This technology was developed out of the need for increased service capacity due to the analog system's limited growth. This technology offers international roaming, high speech quality, increased security, and the capability to develop advanced system features. The development of this technology was completed by a

consortium of 80 pan-European countries working together to provide integrated cellular systems across different borders and cultures.

5. Nordic Mobile Telephone (*NMT*) standard   NMT is the cellular standard that was developed by Sweden, Denmark, Finland, and Norway in 1981. This type of system was designed to operate in the 450MHz and 900MHz frequency bands (these are noted as NMT 450 and NMT 900). NMT systems have also been deployed throughout Europe, Asia, and Australia.

6. Total Access Communications Systems (*TACS*)   TACS is a cellular standard that was derived from the AMPS technology. TACS systems operate in both the 800MHz band and the 900MHz band. The first system of this kind was implemented in England. Later, these systems were installed in Europe, China, Hong Kong, Singapore, and the Middle East. A variation of this standard was implemented in Japan (JTACS).

7. Integrated Dispatch Enhanced Network (*IDEN*)   is the name for an alternative form of cellular communication that operates in the SMR band, just adjacent to the cellular frequency band. IDEN is a blend of wireless interconnect and dispatch services, which makes it very unique when compared to existing cellular and PCS systems. IDEN utilizes a digital radio format called QAM and is a derivative of GSM for the rest of the system, with the exception of the radio link.

Personal Communication Services (PCS)

Personal Communication Services (PCS) is the next generation of wireless communications. PCS is a general name given to wireless systems that have recently been developed out of the need for more capacity and design flexibility than that provided by the initial cellular systems. PCS has similarities to and differences from the current cellular technologies. The similarity between PCS and cellular is the mobility of the users of the service; the differences fall into the applications and spectrum available for PCS operators to provide to the subscribers.

PCS, like its cellular's cousin, is another narrow band service that offers many enhanced data services in conjunction with voice services. PCS was heralded as providing many data services that would enable a person to have one communication device for all of their foreseeable needs. However, due to the bandwidth limitations associated with the PCS systems deployed, the data throughput remained at 9.6kbps.

Wideband PCS has many promises for offering high-speed data, but has not been deployed as of this moment. There are particular problems that must be overcome in order to deploy wideband PCS, and the obvious issue is coexistence with the current PCS system. Coupled with the coexistence problem is the requirement for more base stations due to reduced sensitivity caused by increased bandwidth. The third major problem that needs to be overcome is the offering of subscriber units that can act as dual band units in a vastly diverse PCS marketplace.

The system shown in Figure 1-1, although listed as a cellular system, has the same format and layout as a PCS system. The chief difference is the frequency of operation, which is higher for PCS and requires more base stations in order to cover the same geographic area.

The PCS systems that an operator can utilize are listed as follows for quick reference. It is important to note that in several markets the same operator can and has deployed several types of PCS systems in order to capture market share.

1. *DCS1800*   DCS1800 is a digital standard based on the GSM technology, except that this type of system operates at a higher frequency range: 1800Mhz. The DCS1800 technology is intended for use in PCN type systems. Systems of this type have been installed in Germany and England.

2. *PCS1900*   PCS1900 is the same as DCS1800 and is a GSM system. The only difference between the PCS1900, also called DCS1900, and DCS1800 and GSM is the frequency band of operation. PCS1900 operates in the PCS frequency band for the United States: 1900 MHz.

3. Personal Digital Cellular (*PDC*)   PDC is a digital cellular standard developed by Japan. PDC type systems were designed to operate in the 800MHz band and in the 1.5GHz band.

4. *IS–661*   IS–661 is the technology platform that is being promoted by Omnipoint and is a spread spectrum technology that relies on *time division duplexing* (TDD).

5. *IS–136*   IS–136 is the PCS standard that relies on the NADC system, except it operates in the 1900MHz band.

6. *CDMA*   CDMA is another popular PCS platform that utilizes the same standard as that for CDMA in cellular, except that it operates in the 1900MHz band.

In the United States, PCS operators obtained their spectrum through an action process set up by the *Federal Communications Commission* (FCC). The PCs band was broken into A, B, C, D, E, and F blocks. The A, B, and C blocks involved a total of 30MHz; D, E, and F blocks were allocated 10Mhz.

The spectrum allocation for both cellular and PCS in the United States are shown in Figure 1-2 and Figure 1-3.

It should be noted that the geographic boundaries for PCS licenses are different from those imposed on cellular operators in the United States. Specifically, PCS licenses are defined as MTAs and BTAs. The MTA has several BTAs within its geographic region. There are a total of 93 MTAs and 487 BTAs defined in the United States. Therefore, there are a total of 186 PCS MTA licenses, each with a total of 30MHz of spectrum to utilize. This is in addition to the 1948 BTA licenses awarded in the United States. Regarding the BTA licenses, the C band will have 30MHz of spectrum,

Figure 1-2
U.S. cellular spectrum chart (all frequencies in MHz)

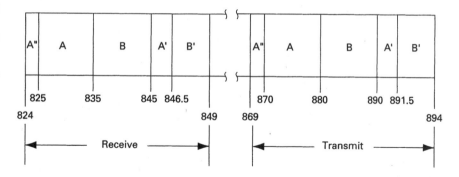

Figure 1-3
PCS spectrum allocation (all frequencies in MHz)

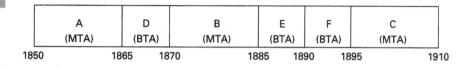

whereas the D, E, and F blocks will have only 10MHz available. All the frequency allocations are duplexed.

WLL

Wireless Local Loop (WLL) utilizes many similar (if not the same) platforms as are used in cellular and PCS systems, and is primarily focused on voice services. WLL, however, is different from cellular or PCS systems in its application, which is fixed and not mobile. Because it is a fixed service, it is often referred to as LMDS or FWPMP. In fact, WLL is often the same as LMDS or FWPMP in its deployment and application. WLL is most applicable in areas where local phone service is not available or cost-effective. Primarily, WLL is a system that connects a subscriber to the local telephone company, PTSN, or PTT by using a radio link as its transport medium instead of copper wires.

There is no specific band that WLL systems occupy or will be deployed in. The systems can either operate in dedicated, protected, spectrum or utilize unlicensed spectrum as their radio access method. Some of the services that fall within the definition of WLL include cordless phone systems, fixed cellular systems, and a variety of proprietary systems.

With the vast choices of system types and spectrum considerations, the choice of which combination to use is directly dependent upon the application and services desired. Some of the additional considerations for choosing the right technology platform involve the determination of the geographic area that needs to be covered, the subscriber density, the usage volume, and patterns expected from the subscribers and the speed of deployment desired.

Because no one radio protocol and service can do everything, the choice of which system to deploy will be driven by the desired market and the applications required to solve a particular set of issues. Some of the more common types of WLL systems involve cellular, PCS, CT−2, and DECT.

WLL has different applications when deployed in a developed or emerging country. For a developed country, the use of WLL takes on the premise of a cordless phone that is an extension of the house phone or PBX. For an emerging country, the choice of WLL is more profound because it is a cheaper alternative to that of laying wire by being more cost-effective as well as quicker to deploy. In many countries, the use of cellular or PCS is quicker, easier, and cheaper to obtain than having a regular landline phone installed.

Figure 1-4 represents a typical WLL system. The WLL system has various nodes that are connected back to a main concentration point. The

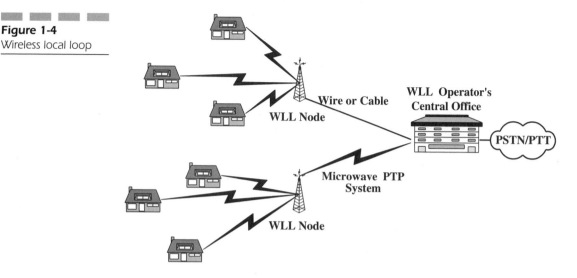

Figure 1-4
Wireless local loop

method for connecting the nodes to the concentration point can be by radio, wire, cable, or a combination of all three.

Table 1-1 is a representation of different technology platforms that can be used, depending on the application involved.

The previous generalization illustrates that there is no single platform or application to use when deploying WLL systems.

GPRS

General Packet Radio Service (GPRS) is an enhancement to GSM type systems that will allow a GSM system to resemble a subscriber's LAN, WAN, and, of course, the Internet. GPRS is an important step in the evolution to 3G, third-generation, mobile telephone networks. The key benefit of GPRS is that it integrates higher throughput packet data to mobile networks, fully enabling mobile Internet applications and a range of other advanced data services.

GPRS gives GSM operators the capability to offer customers better wireless access to the Internet, as well as a wide range of other IP-based services at speeds reaching 115kbps. GPRS enables GSM operators to offer a host of IP services, including email and Web browsing. Of course, it is also an e-commerce enabler from the customer perspective.

Table 1-1

WLL technology
platforms

	WLL Technology
Urban	Digital Cellular
	DECT
	CT2
	LMDS
	Proprietary Radio System
Suburban	Digital Cellular
	DECT
	LMDS
	Proprietary Radio System
Rural	Analog Cellular
	Digital Cellular
	Proprietary Radio Systems

In the quest to provide GPRS, a GSM operator will need to implement various system upgrades and enhancements to enable this technology to be delivered to the customers. GPRS has excellent potential in offering increased data throughput to mobile subscribers. However, GPRS is still a narrow band service and cannot transport efficiently many of the heavy bandwidth applications that a LMDS system can.

LMDS

Local Multipoint Distribution System (LMDS) is a unique, wireless access system whose purpose is to provide broadband access to multiple subscribers in the same geographic area. Point-to-multipoint is a concept in which multiple subscribers can access the same radio platform, utilizing both a multiplexing method and queuing. LMDS, while operating in the microwave frequency band and utilizing similar radio technology as a point-to-point microwave system, enables an operator to handle more subscribers or rather Mbps/km^2 than a microwave point-to-point system using the same amount of radio frequency spectrum.

LMDS utilizes microwave radio as the fundamental transport medium. It is not fundamentally a new technology, but an adaptation of existing technology for a new service implementation. The new service implementation allows multiple users to access the same radio spectrum. LMDS is a wireless system that employs cellular-like design and reuse, except there is no handoff. It can be argued that LMDS is in fact another variant to the WLL portfolio described previously and referenced as proprietary radio systems.

LMDS can be a very cost-effective alternative for a *competitive local exchange carrier* (CLEC). With LMDS, a CLEC can deploy a wireless system without having to experience the heavy capital requirements of laying down cable or copper to reach the customers. The cost effectiveness is born out of the capability to focus the capital infrastructure where the customers are, and at the same time being able to deploy the system in an extremely short period of time.

LMDS consists of two key elements: the physical transport layer and the service layer. The physical transport layer involves both radio and packet/circuit switching platforms. The radio platform consists of a series of base stations that provide the radio communication link between the customers and the main concentration point, usually the central office of the LMDS operator. Figure 1-5 is a high-level system diagram of an LMDS system. The system shown has a layout similar to that for a cellular or PCS mobile system, except that the subscribers are fixed and of course operate at a different frequency.

The system diagram depicts multiple subscribers, customers, surrounding an LMDS hub or base station. The base station is normally configured as a sectorized site for frequency reuse purposes, and there are multiple subscribers assigned to any sector. The amount of channels and the overall frequency plan for the system is driven by the spectrum available in any given market and the amount of capacity required in any geographic zone.

LMDS enables the operator to oversubscribe, as compared to a microwave point-to-point system. Specifically, a single radio channel may have 12Mbps total throughput, but you might be able to offer 24 Mbps or greater for the same channel by allocating it to the entire sector and not specific customers through overbooking. There are QOS issues and specific service delivery requirements with any commercial system. However, the concept of an LMDS system utilizing point-to-multipoint technology can provide vastly greater bandwidth and services to a larger population than a point-to-point system can, utilizing the same spectrum.

Figure 1-5
Generic LMDS system

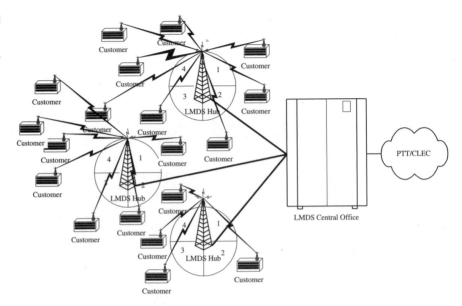

An LMDS system is different from a mobile system in several important ways. The first is that ubiquitous coverage is not required, and in fact is a key advantage. LMDS, if deployed properly, can have the operator provide service only where the customers are actually located, thereby maximizing the capital infrastructure effectiveness and minimizing operating expenses.

The other issue with LMDS relates to the fixed subscriber base that is potentially there. A primary concept with delivering LMDS service is to provide the service not to one customer in a sector, but to multiple customers. The concept is further carried to each building in which the service is deployed. Specifically, LMDS is best positioned when there are multiple customers that utilize the same radio equipment, thereby maximizing the capital deployed at that location.

A brief example of a building having multiple customers is shown in Figure 1-6. The simple concept of having multiple customers per geographic location will minimize the cost of acquisition for any customer, and at the same time reduce operating and capital costs by allowing for some additional concentration. It is important to note that the building where the equipment is to be deployed should be evaluated first to properly establish its bandwidth potential.

In the previous example, it is assumed that access to the wiring closet is achieved for distribution of the services offered. Also, the diagram implies

Figure 1-6
Host location

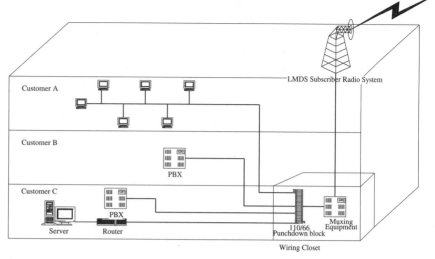

Multiple Customer Location/Dwelling Unit

that there is LOS with the hub site to ensure that the link is of sufficient quality for stable and reliable communication.

LMDS systems can be deployed in a very short period of time (months) and can also be deployed in a method matching exactly where there will be revenue. The deployment involves both the physical transport layer as well as the service layer.

Some of the services that LMDS can offer any given customer or customers are listed as follows for quick reference. It should be noted that the services listed are not all-inclusive of what can be delivered or will be delivered. The only exception is that the service offered cannot have a bandwidth requirement greater than what the radio transport layer can support.

Applications:
- LAN/WAN (VPN)
- T1/E1 Replacement (clear and channelized)
- Fraction T1/E1 (clear and channelized)
- Frame Relay
- Voice Telephony (POTS and enhanced services)
- Video Conferencing
- Internet Connectivity

- Web Services (email, hosting, virtual ISP, etc.)
- E-commerce
- VoIP
- FaxIP
- Long-Distance and International Telephony
- ISDN (BRI and PRI)

The host of services makes an impressive portfolio to offer. Of course, the necessary platforms and connectivity for the network need to be in place to ensure that these services can be offered and effectively delivered. It is interesting that with a LMDS system, as with all other networks, there are on-net and off-net traffic considerations. Ideally, the traffic should be all on-net. When the system initially goes online, however, most if not all the traffic goes off-net and the use of the PTT or another CLEC will be required to facilitate the delivery of the service almost exclusively.

As with all wireless systems, LMDS is no different because there are multiple LMDS systems from which an operator needs to choose to deploy. Some of the system architectures to pick from are FDD, TDD, TDM/ATM, ATM, and FDD/TDM. Coupled with the transport method, the choice of modulation scheme as well as frequency planning options must all be weighed. Additionally, another often-overlooked aspect is the method for actually delivering service to a customer, and the physical and electrical demarcation location and method.

FWPMP

Fixed Wireless Point to Multipoint (FWPMP) is the European variant to LMDS. The equipment deployment and overall concepts are virtually identical to LMDS, but there is a more unified band plan that was implemented in Europe.

Like LMDS, FWPMP consists of two key elements, the physical transport layer and the service layer, and it is a very cost-effective alternative for a CLEC. Using a FWPMP system, a CLEC can deploy a wireless system without having to experience the heavy capital requirements of laying down cable to reach the customers. The cost-effectiveness is born out of the capability to focus the capital infrastructure where the customers are and being able to deploy the system in an extremely short period of time.

FWPMP is effectively a *line of site system* (LOS), which has the same fundamental requirements as a point-to-point microwave system. However, the chief difference between the FWPMP and a PTP system is that multiple subscribers share the same bandwidth, allowing for over subscriber.

There are unique differences, however, in the service offerings allowed for FWPMP in Europe, as compared with LMDS. Some of the chief differences lie in different requirements or restrictions imposed on each of the perspective FWPMP operators. Depending on the country the system is deployed in, the license will have restrictions, imposing a minimum coverage requirement for universal services over a specified period of time (five years). Other differences between LMDS and FWPMP are the spectrum assignments and the channel plans put forth by the European community.

The FWPMP operates in the 26GHz band (24.5–26.5GHz and 27.5–29.5 GHz). It has frequency blocks that are in divisions of 7MHz and the channel allocation is usually done in increments of 28MHz. The channel separation between the duplexed channel blocks for any 7MHz channel grouping is 1008MHz. There is also the 38GHz band that is also being deployed, 37–39.5GHz for FWPMP systems in Europe. LMDS, however, is primarily associated with the 24GHz, 28GHz, and 38GHz spectrum blocks—each having a channel block of 50MHz with four channels that consist of 12.5MHz wide, duplexed.

Two other common bands, sometimes referenced with FWPMP for Europe, are the 3GHz and the 10GHz bands. The 3GHz band covers 3.41–3.6GHz, whereas the 10GHz band covers 10.15–10.65GHz. One of the more obvious issues associated with these bands is the enhanced coverage that a system can have using the lower frequency, as compared with the 26GHz band. However, the bandwidth is not as great with the lower frequencies, and therefore the overall system capacity due to bandwidth as well as frequency reuse issues is not as great. But the cost of deployment for a lower-frequency FWPMP system is also cheaper at the beginning when there is no (or marginal) income.

As with LMDS, the FWPMP band has a host of technology proponents, with the method for implementing the radio system being either TDD, FDD, TDM, ATM, or TDM/FDD (to mention a few of the major variants). The choice of which transport method to use is as daunting as it is with LMDS.

MMDS, MDS, IFTS

Multichannel Multipoint Distribution Systems (MMDS), *Instructional Television Fixed Service* (IFTS), and *Multipoint Distribution Service* (MDS) are all sister bands to LMDS. The combination of MMDS, IFTS, and MDS bands make up what is referred to as wireless cable.

There are a total of 33 channels, each 6MHz wide, which make up the MMDS, MDS, and IFTS bands collectively. The bands, although currently being referenced together, all developed for different reasons. However, the bands were originally broadcast-related because they were one-way oriented. The exceptions were IFTS channels, which have a part of the band allocated for upstream communication.

The MMDS, MDS, IFTS band has numerous subscribers utilizing its service. However there has been increased activity in redefining the services the band can and will offer subscribers. The primary focus of the band is toward high-speed Internet traffic, IP, as compared to video services in conjunction with data. To make this happen, the band(s) have been allocated for two-way communication. But the channels are not paired, as is done commonly in other bands. The two technology types that are competing for use in this band are FDD and TDD systems.

The technologies being deployed for MMDS, MDS, and IFTS bands are similar to LMDS because they involve a sectorized cell site that has multiple subscriber terminals associated with each channel in every sector. One of the key advantages that MMDS, MDS, and IFTS bands have is the frequency these bands operate within. The bands for operation are 2.15–2.162GHz and 2.5–2.686GHz, which do not require strict adherence to line of site for communication reliability or the elimination of rain fade considerations in the link budget.

The chief disadvantage of this band is the coordination that an operator must achieve to utilize a particular frequency in a geographic area. The coordination is exceptionally tricky due to the existence of MMDS, MDS, and IFTS operators that primarily utilize video as their service offering. The coordination issue arises from both upstream and downstream frequency coordination because existing operators designed their systems on a broadcast-system basis.

xDSL

xDSL is the term that is used to describe x-type digital subscriber line technology. There are numerous variants to digital subscriber line technology types, but they all have a similar premise: to convert the access line, twisted pair, into a high-speed data line allowing for a host of services to be offered with the existing infrastructure. DSL technology involves different modulation methods that enhance the data throughput capabilities of an existing access line, local loop.

Just which variant of DSL one may use is entirely dependent upon the application involved or the problem to be solved (see Table 1-2).

LMDS deployments can compete and also use xDSL platforms. The competition with xDSL arises out of the capability to compete against the various bandwidth delivery platforms that LMDS offers to small and medium-size businesses. LMDS can exploit xDSL and use it as a complement for deployment strategy in multiple dwelling buildings where the existing wiring can be exploited, or when reselling the xDSL service is more effective in a geographic area for total service then deploying radio equipment. LMDS can also act as a last mile distribution point for xDSL.

The different types of DSL were listed in the Table 1-2. Of the various forms of DSL used, HDSL and HDSL2 are used for delivering T1/E1 services using existing twisted pairs. SDSL may be used for delivering services in a multiple dwelling unit.

Table 1-2

xDSL

DSL	Data Rate	Comment
HDSL	1.544Mbps or 2.048Mbps	Symmetric, 2 pair
HDSL2	1.544Mbps or 2.048Mbps	Symmetric, 1 pair
SDSL	768kbps	Symmetric, 1 pair
ADSL	1.5–8Mbps-down	Asymmetric, 1 pair
	16–640kbps-up	
RDSL	1.5–8Mbps-down	Asymmetric, 1 pair, but
	16–640kbps-up	changes data rate per
		line condition
CDSL	1Mbps-down	Asymmetric, 1 pair
	16–128kbps-up	
IDSL	64kbps	Symmetric, 1 pair
VDSL	13–52Mbps-down	Asymmetric, 1 pair
	1.5–6Mbps-up	

ADSL is interesting because it is becoming a very popular residential service offering, focusing on the lower end of the data market, the home user. ADSL has multiple business applications that can be used in conjunction with a LMDS operation. One such function involves having an ADSL service used by the microcells used for both cellular/GSM/PCS sites to reduce monthly facility costs. With a partial T1/E1 being used, the bandwidth requirements in both directions fall within the bandwidth capabilities of a residential ADSL service, thereby utilizing a residential service for a commercial application.

WAP

Wireless Application Protocol (WAP) is one of the many broadband protocols being implemented into the wireless arena for the purpose of increasing mobility by giving mobile users the ability to surf the Internet. WAP is being implemented by numerous mobile equipment vendors because it is meant to provide a universal open standard for wireless phones (that is, cellular/GSM and PCS) for the purpose of delivering Internet content and other value-added services. Besides various mobile phones, WAP is also designed for PDAs to also utilize this protocol.

WAP will enable mobile users to surf the Internet in a limited fashion (they can send and receive email and surf the Net in a text format only, without graphics). For WAP to be utilized by a mobile subscriber, the wireless operator, cellular or PCS, needs to implement WAP in their system as well as ensuring that the subscriber units (the phones) are capable of WAP.

The WAP protocol is meant to be used by the following cellular/PCS system types:

■ GSM−900, GSM−1800, GSM−1900

■ CDMA IS−95

■ TDMA IS−136

■ 3G systems—IMT−2000, UMTS, W-CDMA, Wideband IS−95

WAP is fundamentally different from LMDS. Although it delivers wireless data, it does not have the bandwidth to deliver either leased line replacement or support broadband technologies. However, WAP will increase the mobility of many subscribers and enable a host of data applications to be delivered for enhanced services to subscribers.

Bluetooth

Bluetooth is a wireless protocol that operates in the 2.4GHz ISM band. It allows wireless connectivity between mobile phones, PDAs, and other similar devices for the purpose of exchanging information between them. Bluetooth is meant to replace the infrared telemetry portion on mobile phones and PDAs, enabling extended range and flexibility in addition to enhanced services.

Because Bluetooth systems utilize a radio link in the ISM band, there are several key advantages that this transport protocol can exploit. Bluetooth can effectively operate as an extension of a LAN or a peer to peer LAN, and provide connectivity between a mobile device and the following other device types:

- Printers
- PDAs
- Mobile phones
- LCD projectors
- Wireless LAN devices
- Notebooks and Desktop PCs

Some of the key attributes that Bluetooth offers is the range that the system or connection can operate over. Because Bluetooth operates in the 2.4GHz *Industrial-Scientific-Medical* (ISM) band, it has an effective range from 10 to close to 100 meters. The protocol does not require line of site for establishing communication. Its pattern is omnidirectional, which eliminates orientation issues; and can support both ISO and asychronous services, which paves the way for effective use of TCP/IP communication.

Bluetooth is different from LMDS and WAP, but again it looks at delivering data connectivity over radio. Bluetooth is also different from LMDS because of the applications, use of the unlicensed band, and focus on end user devices. Bluetooth is meant to be a LAN extension that fosters communication connection, not delivers bandwidth.

Typical Central Office (CO)

A *central office* (CO) is anything but typical. Although a CO typically delivers voice service and provides the local loop aspect for telephony, the particular functions and services the CO can offer and deliver are extremely varied. For example, the primary services the CO would deliver in a residential area are voice services. In a business district, however, the CO may be more structured to support data and centrex type services. With the advent of Internet popularity, many residential COs that primarily delivered voice services are now transitioning from a circuit switched system to one of a packet switching system.

A simplified example of a typical CO layout is shown in Figure 1-7. Naturally, the dimensions and specific equipment required for the facility will need to factor in the type of services to be provided, as well as the timeframe the design is to encompass (growth).

The following is a brief list of the elements in a fixed network design associated with a wireless system. There are obviously more elements to a network design, including coordination issues with adjacent markets as well as with the various vendors.

Typically, a CO consists of an equipment room, toll room, power room, and operations room. The functions of each are unique because the equipment room has the switching and packet platforms for treating and servicing the subscribers' needs. The toll room, also referred to as the interconnect and telco room, is the area of the MSC when the system interfaces to the PSTN, CLECs, IXCs, and other outside carriers. The purpose of the toll room is to

Figure 1-7
Typical CO layout

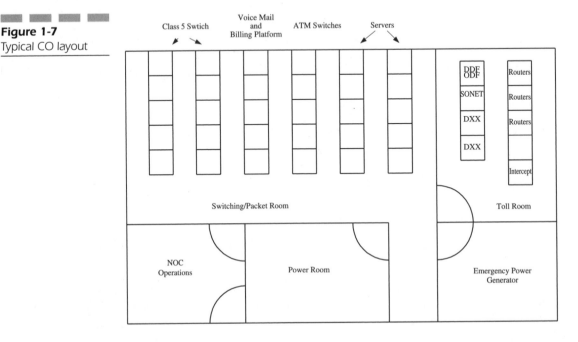

provide the portal for entry and exit of services for the CO. The power room usually houses the rectifiers, batteries, and generator for emergency backup purposes. The operations center is the area where the craft personnel perform the data entry, and monitoring and maintenance of the network itself.

The following list of topics should prove helpful in establishing the resources and timing needed for a fixed network design to be successful.

Equipment Room

- Class 5 Switch
- ATM Switches
- Voice Mail system
- Servers
- Billing System

Toll Room

- STP
- DXX Equipment
- Routers
- Intercept Equipment

The previous list does not address the issue of colocation with other service providers and the need to create a separate area for the operator to maintain and upgrade equipment.

CFR47

The Code of Federal Regulations, Title 47, CFR47, contains the rules and regulations put forth by the FCC for operation of all telecommunications in the United States. The rules are broken down into multiple parts and each part applies to a specific function. It is important to note that rules within a particular part can and often do reference other parts of CFR47.

CFR47 sets the requirements and rules at the Federal level. The Public Utility Commission in each state, or its name variant, also imposes rules regarding tariffs that need to be followed. Even with the enactment of the telecommunication act, local zoning rules for construction of facilities have not been abolished. The telecommunication act clarified some of the local governments dos and don'ts regarding antenna-siting issues.

The current CRF47 can be obtained either through the Government Printing Office or via the Web. The rules are continuously being modified, and it is important to be aware of and respond to changes with comments either for or against various proposed rule changes. A good place to start looking for information on any topic is the FCC Web site: `www.fcc.gov`.

Table 1-3 shows the current breakdown of the various parts of CFR47. A key point is that the parts are not contiguous because there are numerous parts held in reserve for future use. This is why the rules have gaps between the various numberings. The comment portion contains a very brief listing of some of the key services that are governed primarily by that part of CFR47.

VARs and Their Role for Success

For the successful deployment of an LMDS or any wireless system, the use of VARs is essential for both the acquisition and retention of customers. Because there are many types of VARs, or company types in which the term VAR is attached to, the first step an operator has to decide what business they are in. Specifically, the question must be put forth for an LMDS operator to determine whether they are a wholesale pipe provider only, LAN/WAN full-service provider, wholesale and retail bandwidth provider, or some variant of those.

Once the decision is made, which can be almost an elusive target, the decision must be made about how and what type of VARs will contribute to the sales and operation of the company. The choice of how VARs interact with the company can and will change over time as market conditions change and different issues arise as the company matures or merges with other companies.

There can also be several types of VARs that make up the complement of service enhancements for the LMDS operator. For instance, one type of VAR can be focused on reselling bandwidth where the LMDS operator provides the physical equipment to a building or multiple-dwelling unit. The look and feel of the service the customers sees is the VAR's and not the LMDS operators.

Another aspect for a VAR is one that installs the wiring connecting the LMDS equipment to the customer in each location. The VAR can be a SOHO LAN company or an Enterprise LAN company. The primary issue here is that the VAR can also be directly supporting customers' LAN and IT departments, either as an adjunct service or as a complete outsourcing package. The advantage of having this type of VAR in the product-offering

Table 1-3

CFR47

Part	Name	Comments
0	Commission Organization	General
1	Practice and Procedure	EMF
2	Frequency Allocations and Radio Treaty Matters; General Rules and Regulations	
3	Authorization and Administration of Accounting Authorities in Maritime and Maritime Mobile-Satellite Radio Services	
5	Experimental Radio Services (Other than Broadcast)	
11	*Emergency Alert System* (EAS)	
13	Commercial Radio Operators	
15	Radio Frequency Devices	Unlicensed Devices
17	Construction, Marking and Lighting of Antenna Structures	Towers, FAA
18	*Industrial, Scientific, and Medical Equipment* (ISM)	
19	Employee Responsibilities and Conduct	
20	Commercial Mobile Radio Services	
21	Domestic Public Fixed Radio Services	
22	Public Mobile Services	Cellular
23	International Fixed Public Radiocommunication Services	
24	*Personal Communication Services* (PCS)	
25	Satellite Communications	
26	*General Wireless Communication Services* (GWCS)	
27	*Wireless Communication Service* (WCS)	
32	Uniform System of Accounts for Telecommunication Companies	
36	Jurisdictional Separations Procedures; Standard Procedures for Separating Telecommunication Property Costs, Revenues, Expenses, Taxes, and Reserves	
41	Telegraph and Telephone Franks	
42	Preservation of Records of Communication Common Carriers	
43	Reports of Communication Common Carriers and Certain Affiliates	
51	Interconnection	
52	Numbering	
53	Special Provision Concerning Bell Operating Companies	
54	Universal Service	
59	Infrastructure Sharing	
61	Tariffs	
62	Applications to Hold Interlocking Directorates	
63	Extension of Lines and Discontinuance, Reduction, Outage, and Impairment of Service by Common Carriers	
65	Interstate Rate of Return Prescription Procedures and Methodologies	
68	Connection of Terminal Equipment to the Telephone Network	
69	Access Charges	
73	Radio Broadcast Services	
74	Experimental Radio, Auxiliary, Special Broadcast, and other Program Distributional Services	
76	Cable Television Service	
78	Cable Television Relay Service	
79	Closed Captioning of Video Programming	
80	Stations in the Maritime Services	
87	Aviation Services	
90	Private Land Mobile Radio Services	SMR
95	Personal Radio Services	
97	Amateur Radio Service	
100	Direct Broadcast Satellite Service	
101	Fixed Microwave Services	LMDS, PTP

mix is that it can elevate the issue of absorbing the LAN/IT problems of a customer into the operation of the LMDS system.

When selecting VARs to assist in the deployment, sales, and retention of customers, there are some salient questions that should be asked, listed as follows:

- What services do you want them to do?
- How do you want them to represent the company?
- Do they have the qualifications to perform the task desired?
- How long have they been doing this type of work?
- What is the personnel mix in terms of experience relative to the desired service offered?
- How many trained personnel do they have versus what you expect the need is (realistically)?
- What do you expect the VAR to do for ongoing maintenance and service?
- How do you plan on compensating them for initial sales, added services, and, of course, customer retention?
- Does a background check on the firm produce any concerning issues (D&B, references, etc.)?
- Who in the LMDS company will be responsible for the VAR relationship?

The previous list is not all-inclusive, but if you can answer the questions, the chance for a misstep with them is minimized. It cannot be overstressed that the VAR is what the customer interacts with, not the LMDS company, until there is a problem. Often, in the race to get service online there is great pressure to have the apparent installation and distribution issues solved by selecting various VARs. However, because no two companies or locations are the same, it is important to ensure that the VARs used will properly represent the company in both presence as well as technical competency.

References

Barron, Tim. "Wireless Links for PCS and Cellular Networks." *Cellular Integration*, September 1995, 20–23.

Fulton, Fiona. "GPRS' Next Challenge: Billing." *Telecommunications*, December 1999, 65–66.

Goralski, Walter. *ADSL and DSL Technologies*, New York, NY: McGraw-Hill, 1998.

Lynch, Dick. "Developing a Cellular/PCS National Seamless Network." *Cellular Integration*, September 1995, 24–26.

McClelland, Stephen. "Europe's Wireless Futures." *Microwave Journal*, September 1999, 78–107.

Rusch, Roger. "The Market and Proposed Systems for Satellite Communications." *Applied Microwave & Wireless*, Fall 1995, 10–34.

Smith, Clint and Curt Gervelis. *Cellular System Design and Optimization*. New York, NY: McGraw-Hill, 1996.

Smith, Clint. *Practical Cellular and PCS Design*. New York, NY: McGraw-Hill, 1997.

Winch, Robert G. *Telecommunication Transmission Systems*, 2nd edition. New York, NY: McGraw Hill, 1998.

Technology

Introduction

There is a plethora of technology issues that a LMDS operator needs to factor into the design and ongoing operation of the system. The technology decisions are important also for decisions involving the implementation of various services and functions from the LMDS operators platform. Technology decisions also are involved when evaluating competitive service offerings and for the determination of niche markets, features, and services that can be used for the LMDS operators' advantage.

This chapter attempts to address many of the technology issues that a LMDS operator is often confronted with and will need to address when making decisions regarding which technology or technologies to utilize for a LMDS, MDS or FWPMP system. The technology topics will be covered in a depth that will provide general understanding of the various commonality and differences between the vast array of platforms that have to be used.

In particular parts of any LMDS network, the PTT/CLEC can be a competitor, service provider, partner, customer, or all of them combined. In many cases, the LMDS operator can complement a local PTT/CLEC or ISP in the service delivery. In fact, most PTT/CLECs utilize a variety of transport platforms to deliver specific services to a diverse customer base.

Analog

The reference to analog in wireless can and does takes on a multitude of meanings that invoke different responses, depending on the situation and system it is applied or referenced to.

Analog communication references any communication that does not utilize a digital modulation format to convey its information—voice. Specifically, a form of analog communication is the AM or FM station that you listen to in your vehicle or home. Typically, the term "analog" usually refers to an FM-modulated signal. When people reference analog channels in a cellular communication system, they are referring to the 30kHz AMPS channel that was and is used prior to the advent of digital radio platforms. It must be stressed that although voice is being transported as analog, the system will utilize digital modulation for conveying control and subscriber information.

For a PTT/CLEC, analog is typically the local loop where voice is delivered over copper wire from the wiring center to the residence. This is important for an LMDS operator if the service offering involves voice, POTS, or interfaces to a PBX that has an analog line card, to mention two quick examples.

Digital

Digital or digital modulation is prevalent throughout the entire wireless industry. Digital communication references any communication that utilizes a modulation format that relies on sending the information in any type of data format. More specifically, digital communication is where the sending location digitizes the voice communication and then modulates it. At the receiver, the exact opposite is done.

Data is digital, but it needs to be converted into another medium in order to transport it from point A to point B; more specifically, between the base station and the host terminal. The data between the base station and the host terminal is converted from a digital signal into an RF energy whose modulation is a representation of the digital information that enables the receiving device, base station, or host terminal to properly replicate the data.

Digital radio technology is deployed in a cellular/PCS/SMR, and specifically in an LMDS system to increase the quality and capacity of the wireless system over its analog counterpart. The use of digital-modulation techniques enables the wireless system to transport more bit/Hz than would be possible with analog signaling utilizing the same bandwidth. There are, of course, many different digital modulation schemes utilized in LMDS, but the more common ones are the following:

- QPSK
- 4QAM
- 16QAM
- 64QAM

The data rate that each can support increased with modulation complexity, but there is a tradeoff that is made. With increased modulation complexity, the range of the site is reduced, resulting in higher capital deployment costs at the beginning of the systems life cycle.

For LMDS, there are several competing digital techniques that have been or are being deployed. Each of the technologies has advantages and disadvantages that need to be understood prior to their implementation.

However, the major benefits associated with utilizing digital radios for an LMDS environment involve the following key topics:

- Increased capacity over analog
- Reduced capital infrastructure costs for bits/Hz
- Reduce the capital per subscriber cost through overbooking
- Features both present and future
- Encryption

Mobile Wireless Systems

There are numerous mobile wireless systems that have been deployed throughout the world. Each of the wireless mobility systems has unique advantages and disadvantages. LMDS and FWPMP have often been referenced as cellular-like. Although there are many similarities, there are also significant differences. One in particular is that the subscribers are stationary and coverage does not have to be ubiquitous.

Table 2-1 represents the most popular wireless mobility service offerings that have been deployed.

Table 2-1

Cellular systems

	AMPS	NAMPS	TACS	NMT450	NMT900	C450
Base Tx MHz	869–894	869–894	935–960	463–468	935–960	461–466
Base Rx MHz	824–849	824–849	890–915	453–458	890–915	451–456
Multiple Access Method	FDMA	FDMA	FDMA	FDMA	FDMA	FDMA
Modulation	FM	FM	FM	FM	FM	FM
Radio Channel Spacing	30kHz	10kHz	25kHz	25kHz	12.5kHz	20kHz (b) 10kHz (m)
Number Channels	832	2496	1000	200	1999	222 (b) 444 (m)
CODEC	NA	NA	NA	NA	NA	NA
Spectrum Allocation	50MHz	50MHz	50MHz	10MHz	50MHz	10MHz

Table 2-2

Digital cellular systems

	IS-136	IS-136*	IS-95	GSM	iDEN
Base Tx MHz	869–894	851–866	869–894	925–960	851–866
Base Rx MHz	824–849	806–821	869–894	880–915	806–821
Multiple Access Method	TDMA/FDMA	TDMA	CDMA/FDMA	TDMA/FDMA	TDMA
Modulation	Pi/4DPSK	Pi/4DPSK	QPSK	0.3GMSK	16QAM
Radio Channel Spacing	30kHz	30kHz	1.25MHz	200kHz	25kHz
Users/Channel	3	3	64	8	3/6
Number Channels	832	600	9 (A) 10 (B)	124	600
CODEC	ACELP/VCELP	ACELP	CELP	RELP-LTP	
Spectrum Allocation	50MHz	30MHz	50MHz	50MHz	30MHz

*Down banded IS-136 in the SMR band

Table 2-2 represents various digital cellular systems that are used throughout the world.

Table 2-3 represents various PCS systems that are used throughout the world and in the United States in particular.

AMPS

Avanced Mobile Phone System (AMPS) is the cellular standard that was developed for use in North America. This type of system operates in the 800Mhz frequency band. AMPS systems have also been deployed in South America, Asia, and Russia.

A typical AMPS system is shown in Figure 2-1 for reference.

TACS

Total Access Communications Systems (TACS) is a cellular standard that was derived from the AMPS technology. TACS systems operate in both the 800MHz band and the 900MHz band, but with 25kHz channels versus the

Table 2-3

PCS systems

	IS-136	IS-95	DCS1800	DCS1900	IS661
Base Tx MHz	1930–1990	1930–1990	1805–1880	1930–1990	1930–1990
Base Rx MHz	1850–1910	1850–1910	1710–1785	1850–1910	1850–1910
Multiple Access Method	TDMA/FDMA	CDMA/FDMA	TDMA/FDMA	TDMA/FDMA	TDD
Modulation	Pi/4DPSK	QPSK	0.3GMSK	0.3GMSK	QPSK
Radio Channel Spacing	30kHz	1.25MHz	200kHz	200kHz	5MHz
Users/Channel	3	64	8	8	64
Number Channels	166/332 /498	4–12	325	25/50 /75	2–6
CODEC	ACELP/VCELP	CELP	RELP-LTP	RELP-LTP	CELP
Spectrum Allocation	10/20 /30Mhz	10/20 /30Mhz	150MHz	10/20 /30Mhz	10/20 /30Mhz

Figure 2-1

Typical AMPS cellular system

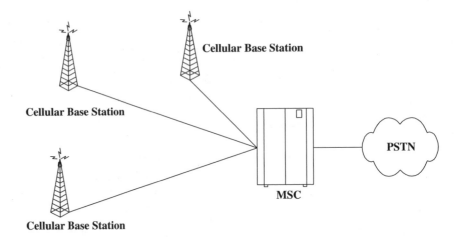

30kHz channels for AMPS. The first system of this kind was implemented in England. Later, these systems were installed in Europe, China, Hong Kong, Singapore, and the Middle East. A variation of this standard, JTACS, was implemented in Japan and is very similar to TACS except that the channel bandwidth is smaller: 12.5kHz.

NAMPS

Narrow Band AMPS (NAMPS) is a product that is used in parts of the United States, Latin America, and other parts of the world. Specifically, NAMPS is an analog radio system that is very similar to AMPS, except that it utilizes 10kHz-wide voice channels instead of the standard 30kHz channels. The obvious advantage with this technology is the capability to deliver, under ideal conditions, three times more capacity than that of regular AMPS.

NAMPS is able to achieve this smaller bandwidth through changing the format and methodology for SAT and control communications from the cell site to the subscriber unit. In particular, they use a sub-carrier method and a digital color code in place of SAT. These two methods make it possible to use less spectrum while communicating the same amount or even more information while increasing the capacity of the system with the same spectrum.

Of course, this advantage in capacity requires a separate transmitter, either PA or transceiver, for each NAMPS channel deployed. However, the control channel that is used for the cell site is the standard control channel, 30kHz, used by AMPS and other technology platforms for cellular communication. Figure 2-2 shows a simple diagram of a NAMPS system. The C/I requirements differ from a regular AMPS system, which has a direct impact on the capacity of the system, due to the narrower bandwidth channels.

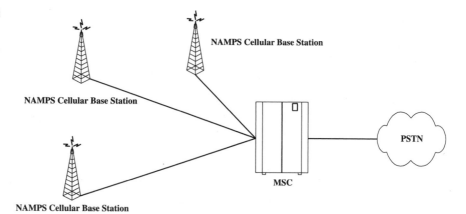

Figure 2-2
NAMPS system

NAMPS Cellular Base Station

NAMPS Cellular Base Station

NAMPS Cellular Base Station

PSTN

MSC

IS-136

IS-136, also known as IS-54, is the digital cellular standard developed in the United States using TDMA technology. Systems of this type operate in the same band as the AMPS systems and are also used in the PCS spectrum. IS-136, therefore, applies to both the cellular and PCS bands. In some unique situations, it also applies to down-banded IS-136, which operates in the SMR band.

IS-136 utilizes TDMA technology that allows multiple users to occupy the same channel through the use of time division. The TDMA format utilized in the United States is also referred to as *the North American Dual Mode Cellular* (NADC). IS-136 is an evolution of the IS-54 standard and enables a feature-rich technology platform to be utilized by the current cellular operators.

TDMA, which utilizes the IS-136 standard, is currently deployed by several cellular operators in the United States. IS-136 utilizes the same channel bandwidth as the analog cellular, 30kHz per physical radio channel. However, IS-136 enables three and possibly six users to operate on the same physical radio channel at the same time. The IS-136 channel presents a total of six time slots in the forward and reverse direction. At present, IS-136 utilizes two time slots per subscriber, with the potential to go to half-rate vocoders that require the use of only one time slot per subscriber.

IS-136 has many advantages in its deployment in a cellular system, including the following:

- Increased system capacity—up to three times over analog
- Improved protection for adjacent channel interference
- Authentication
- Voice privacy
- Reduced infrastructure capital to deploy
- Frequency plan integration over CDMA
- Short message paging

Integrating IS-136 into an existing cellular system can be done more easily than when deploying CDMA. The use of IS-136 in a network requires the use of a guardband to protect the analog system from the IS-136 signal. However, the guardband required consists of only a single channel on either side of the spectrum block allocated for IS-136 use. Depending on the actual location of the IS-136 channels in the operator's spectrum, it is possible to require only one guardband channel (or no guardband channel).

The IS-136 has the unique advantage of affording the implementation of digital technology into a network without elaborate engineering requirements. The implementation advantages mentioned for IS-136 also facilitate the rapid deployment of this technology into an existing network.

The implementation of IS-136 is further augmented by requiring only one channel per frequency group as part of the initial system offering. The advantage with requiring only one channel per sector in the initial deployment is the minimization of capacity reduction for the existing analog network. Another advantage with deploying one IS-136 channel per sector is that it initially eliminates the need to preload the subscriber base with dual-mode, IS-136 handsets.

Because IS-136 operates in both the cellular and PCS bands, one of the advantages that can exploited with IS-136 is the capability to roam and aggregate system capacity for an operator who has both cellular and PCS spectrum.

Finally, IS-136 systems are also used for wireless local loop systems.

iDEN

The iDEN system involves integrating mobile phone and dispatch technologies together. The services that are integrated into iDEN involve dispatch, full duplex telephone interconnect, data transport, and short messaging services.

The dispatch system involves a feature called group call, in which several people can engage in a conference. The user list is preprogrammed and the conference call can be set up just as it is done in two-way or SMR, except that the connection can take place utilizing any of the frequencies that are available from the pool of channels where the subscriber is physically located.

The telephone interconnect and data transport are meant to offer conventional mobile communications. The short messaging service enables the iDEN phones to receive up to 140 characters for an alphanumeric message.

CDMA

CDMA is a spread spectrum technology platform that enables multiple users to occupy the same radio channel and frequency spectrum at the same time. CDMA is being utilized for microwave point-to-point communication and satellite communication, and is also used by the military. With

CDMA, each of the subscribers or users utilize their own unique code to differentiate themselves from other users. CDMA offers many unique features, including the capability to thwart interference and improved immunity from multipath effects due to its bandwidth.

The benefits associated with CDMA are as follows:

- Increased system capacity over analog and TDMA
- Improved interference protection
- No frequency planning required between CDMA channels
- Improved hand-offs with MAHO and soft hand-offs
- Fraud protection due to encryption and authentication
- Accommodation of new wireless features

CDMA, a wideband spread spectrum technology, is based on the principle of *direct sequence* (DS). The CDMA channel utilized is reused in every cell of the system and is differentiated by the pseudo random number (PN) code that it utilizes.

Despite the apparent advantages of CDMA for a cellular system, there are several implementation concerns regarding deploying it into an existing system. The introduction of CDMA into an existing AMPS system will require the establishment of a guardband and *guardzone*. The guardband and guardzone are required for CDMA to ensure that the interference received from the AMPS system does not negatively impact the capability for CDMA to perform well.

GSM

Global System for Mobile Communications (GSM) is the European standard for digital cellular systems operating in the 900MHz and 1800MHz bands. This technology was developed out of the need for increased service capacity due to the analog systems' limited growth. This technology offers international roaming, high speech quality, increased security, and the capability to develop advanced system features. The development of this technology was completed by a consortium of 80 pan-European countries, working together to provide integrated cellular systems across different borders and cultures.

GSM is a European Standard that has achieved worldwide success. GSM has many unique features and attributes that make it an excellent digital radio standard to utilize. GSM has the unique advantage of being the most widely accepted radio communication standard at this time.

GSM consists of the following major building blocks: the *Switching System* (SS), the *Base Station System* (BSS), and the *Operations and Support System* (OSS). The fundamental building blocks for a GSM system are shown in Figure 2-3.

The GSM system shown has several sub-components associated with each of the three major elements. The first is the *Base Station System* (BSS) which comprises both the *Base Station Controller* (BSC) and *the Base Transceiver Stations* (BTS). In an ordinary configuration, several BTSs are connected to a BSC and then several BSCs are connected to the MSC. The BTS is the radio base station that contains the physical base station radios that are used to communicate with the mobile subscriber units. The BTSs are arranged in a star configuration. The BTSs are connected to the BSC via a physical link referred to as the Abis link.

The BSC is functionally a high-capacity switch that is responsible for all the radio-related functions, such as the management of the radio resources,

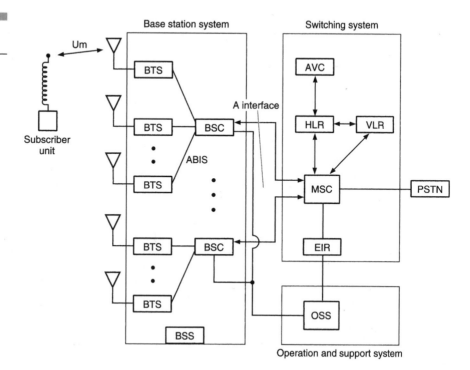

Figure 2-3
GSM system

BTS parameters, and handover functions (to mention but a few). The BSC can either be centrally located with the MSC or it can be distributed throughout the entire network, depending on the interconnect arrangement desired for the network. The BSC communicates to the MSC via the A link interface, which is an open interface. The A link enables different BSC manufacturers to interconnect with different MSC manufacturers in the same market without having to resort to an IS-41 type interface.

The *Switching System* (SS) consists of five main components: *Mobile Switching Center* (MSC), *Home Location Register* (HLR), *Visitor Location Register* (VLR), *Authentication Center* (AUC), and *Equipment Identity Register* (EIR).

The MSC's role in the switching system is to perform all the telephony-switching functions for the network. The MSC coordinates all the traffic between the PSTN and the mobile system. Some of the functions the MSC is responsible for include SS7, network interface, and switching.

The HLR stores information about each of the subscribers that are resident to it. Each system has at least one HLR and it contains permanent subscriber information such as the type of subscriber and the features they subscribe to. The HLR is an integral part of the MSC, but is a separate node to the MSC itself.

The VLR, which is similar to the HLR, is a database that contains information about subscribers that the MSC needs to properly treat each of the calls that are placed by the subscriber. The VLR is utilized by the MSC when the subscriber is not resident to the HLR that is a node to the MSC.

The AUC is another node in the Switching System, except it directly supports the HLR and provides the authentication and encryption parameters required for each call placed on a GSM system.

The EIR is a database that maintains a variety of lists that show faulty and stolen equipment. The subscriber equipment is checked as part of the call treatment for the system.

The third major component to the GSM system involves the *Operations and Support System* (OSS), which is the main interface for operations personnel to monitor and take corrective action to the network. The OSS is also the platform that is used to make any additions or deletions to the network's configuration.

DCS1900 and DCS1800

DSC1800 and DCS 1900 are both GSM systems. The difference between DCS 1800 and DCS 1900 is the frequency of operation. DCS 1800 operates in the 1800 MHz band and is normally associated with European systems. DCS 1900 is a GSM system that operates in the U.S. wideband PCS frequency band at 1900MHz.

The components that make up either a DCS 1800 or a DCS 1900 system are the same as those of a GSM system, except that the frequency of operation has been upbanded from the 900MHz band to the 1800- and 1900MHz bands.

DCS1800 and DCS1900 are also referred to as PCS (PCS1800 and PCS1900).

PCS

Personal Communication Systems (PCS) are systems or services that have recently developed out of the need for more capacity and design flexibility than that provided by the initial cellular systems.

PCS is the next generation of wireless communications and has similarities to and differences from the current cellular technologies. The similarity between PCS and cellular is the mobility of the user of the service. The differences between PCS and cellular fall are the applications and spectrum available for PCS operators to provide to the subscribers.

PCS spectrum in the United States allocation is shown in Figure 2-4.

The geographic boundaries for PCS licenses are different from those for cellular operators in the United States. Specifically, PCS licenses are defined as MTAs and BTAs. The MTA has several BTAs within its geographic region. There is a total of 93 MTAs and 487 BTAs defined in the United States. Therefore, there is a total of 186 MTA licenses awarded for

Figure 2-4
(a) Cellular (b) PCS
spectrum allocations

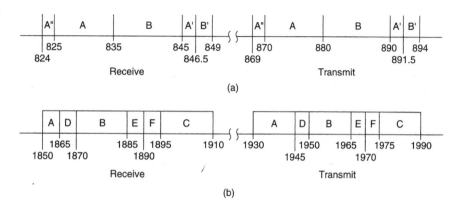

the construction of a PCS network, each with a total of 30MHz of spectrum to utilize. In addition, there are a total of 1948 BTA licenses awarded in the United States. Of the BTA licenses, the C band will have 30MHz of spectrum, whereas the D, E, and F blocks will have only 10MHz available.

Currently, there is no one standard for PCS operators to utilize for picking a technology platform for their networks. The choice of PCS standards is daunting, and each has its own advantages and disadvantages associated with it. The current philosophy in the United States is to let the market decide which standards are the best. This is significantly different from that used for cellular, in which every operator has one set interface for the analog system to operate from.

There are a few major standards that have been picked by the licensees for the A and B PCS operators. The standards so far selected for PCS are DCS-1900, IS-95, IS-661, and IS-136. DCS-1900 utilizes a GSM format and is an upbanded DCS-1800 system. IS-95 is the CDMA standard that will be utilized by cellular operators, except upbanded to the PCS spectrum. IS-661 is a Time Division Duplex system offered by Omnipoint Communications. The IS-136 standard is an upbanded cellular TDMA system that is currently being used by cellular operators.

CDPD

CDPD is a packetized data service that uses its own air interface standard. The CDPD system utilized by the cellular operators is functionally a separate data communication service that physically shares the cell site and cellular spectrum.

CDPD has many applications, but most are applicable for short bursty-type data applications, not large file transfers. CDPD applications of short messages include email, telemetry applications, credit card validations, and global positioning.

CDPD does not establish a direct connection between the host and server locations. Instead, it relies on the OSI model for packet switching data communications, which routes the packet data throughout the network. The CDPD network has various layers: layer 1 is the physical layer, layer 2 is the data link itself, and layer 3 is the network portion of the architecture. CDPD utilizes an open architecture and has incorporated authentication and encryption technology into its airlink standard.

The CDPD system consists of several major components. A block diagram of a CDPD system is shown in Figure 2-5.

▬▬ ▬▬ ▬▬ ▬▬

Figure 2-5
CDPD

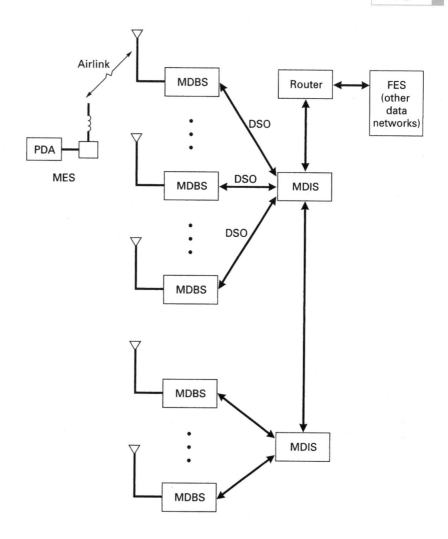

The *Mobile End System* (MES) is a portable, wireless computing device that moves around the CDPD network, communicating with the MDBS. The MES is typically a laptop computer or other personal data device that has a cellular modem.

The *Mobile Data Base Station* (MDBS) resides in the cell site itself and can utilize some of the same infrastructure that the cellular system has for transmitting and receiving packet data. The MDBS acts as the interface between the MES and the MDIS. One MDBS can control several physical radio channels, depending on the site's configuration and loading requirements. The MDBS communicates to the MDIS via a 56Kbps data link.

Often, the data link between the MDBS and MDIS utilizes the same facilities as the cellular system, but occupies a dedicated time slot.

The *Mobile Data Intermediate System* (MDIS) performs all the routing functions for CDPD. It performs these tasks by utilizing the knowledge of where the MES is physically located within the network itself. Several MDISs can be networked together to expand a CDPD network.

The MDIS also is connected to a router or gateway that connects the MDIS to a *Fixed End System* (FES). The FES is a communication system that handles layer 4 and higher transport functions.

The CDPD system utilizes a *Gaussian minimum-shift keying* (GMSK) method of modulation and can transfer packetized data at a rate of 19.2kbps over the 30kHz wide cellular channel. The frequency assignments for CDPD can take on two distinct forms. The first form of frequency assignment dedicates specific cellular radio channels to be utilized by the CDPD network for delivering the data service. The other method of frequency assignment for CDPD utilizes channel hopping, where the CDPD's *Mobile Data Base Station* (MDBS) utilizes unused channels for delivering its packets of data. Both methods of frequency assignments have advantages and disadvantages.

Utilizing a dedicated channel assignment for CDPD has the advantage of not having the CDPD system interfere with the cellular system it is sharing the spectrum with. By enabling the CDPD system to operate on its own set of dedicated channels, there is no real interaction between the packet data network and the cellular voice network. However, the dedicated channel method reduces the overall capacity of the network so this might not be a viable alternative, depending on the system-loading conditions.

If the method of channel hopping is utilized for CDPD and it is part of the CDPD specification, the MDBS for that cell or sector will utilize idle channels for the transmission and reception of data packets. If the channel that is being used for packet data is assigned by the cellular system for a voice communication call, the CDPD MDBS detects the channel's assignment and instructs the MES to retune to another channel before it interferes with the cellular channel. The MDBS utilizes a scanning receiver or sniffer, which scans all the channels it is programmed to scan to determine which channels are idle and which are in use.

The disadvantage of the channel-hopping method involves the potential interference problem with the cellular system. Coexisting on the same channels with the cellular system can create mobile to base station interference, which occurs because of the different hand-off boundaries for CDPD and cellular for the same physical channel. The difference in hand-off boundaries occurs because CDPD utilizes a BER for hand-off determination and the cellular system utilizes RSSI at either the cell site, analog, or MAHO for digital.

GPS

Global Positioning Satellite (GPS) is a series of 24 satellites that orbit the earth to provide position and timing information. The GSP satellite system consists of a total of 24 satellites over six regions of the world, ensuring that at least four satellites are visible anywhere in the world. (Visibility, of course, is dependent upon the lack of obstructions near the ground.)

GPS is used in LMDS systems for synchronization purposes by providing a stable clock or primary reference source for the various components of the network to extract timing from. The GPS timing source is commonly used for a stratum 1E primary reference source.

GPS satellites transmit on two frequencies using a spread spectrum modulation technique, CDMA, using BPSK. The GSP satellite frequencies are as follows:

L1 = 1575.42MHz

L2 = 1227.60MHz

The GPS receiver sensitivity (or the minimal required signal level for proper detection) is directly dependent upon the frequency received and the code that is used:

L1 C/A code = −160 dBW

L1 P code = −163 dBW

L2 P code = −166 dBW

GSP is used not only for Central Office timing purposes, but also for iDEN and CDMA systems for the call processing and signaling used between the cell sites for handoffs.

Differential GPS is used to increase the positional accuracy over a regular GSP receiver. Differential GSP will improve the position accuracy so that the position read using a differential GSP is within 10 meters of the actual position. Ordinary GSP, however, can place accuracy within only 100 meters of the position. Therefore, the advantages of using differential GPS for wireless surveying purposes is clear.

Stratum Clocking

A stratum clock has four levels of stratum. Each of the four levels, 1 through 4, represent different timing requirements and the levels are directly associated with the type of electronics they are applied to in the telecommunication industry. The hierarchy is such that stratum 1 is the highest and stratum 4 is the lowest level of the stratum clock systems.

Stratum 1 is the top or the highest level in the stratum hierarchy. The stratum 1E utilizes typical sources for timing as GSP and LORAN. The Stratum 1 clock utilizes a traceable timing reference and has the highest hold over period because when it loses its source, it will keep its accuracy for the longest period of time of all the stratum levels. Stratum 1 is referred to as the primary reference source for the system, like LMDS, and is required to ensure that proper timing for all the data communication is done correctly.

Stratum 2 is the second-highest level in the stratum hierarchy. Stratum 2 clocks are normally found in tandems and STPs.

Stratum 3 is the third-highest level in the stratum hierarchy and is normally found associated with DACS and digital switches utilized for wireless and landline applications.

Stratum 4 is the fourth-highest level in the stratum hierarchy (the lowest stratum level) and is normally associated with a PBX or similar network element.

dBi and dBd

The difference between dBi and dBd is 2.14 dB. Specifically, a dBi is a reference to a theoretical isotropic antenna that radiates equally in all directions and does not exist. The dBd is a reference to a dipole antenna that has 2.14 dB of gain over an isotropic antenna:

$$dBd = 2.14 + dBi$$

To convert a value in dBi to the equivalent dBd value, the following equation is utilized:

$$dBd = dBi - 2.14$$

Table 2-4 can be used to help reinforce the conversion process that shows the calculated values along with the nearest approximate value found.

Table 2-4

dBi to dBd

dBi	dBd
5	2.86 (3 dBd)
10	7.86 (8 dBd)
12	9.86 (10 dBd)
14	11.86 (12 dBd)
18	15.86 (16 dBd)
21	18.86 (19 dBd)

Table 2-5

dBd to dBi

dBd	dBi
3	5.14
10	12.14
12	14.14
14	16.14
18	20.14
21	23.14

Often, antenna manufacturers will present an antenna gain in dBd, and you will need to convert the value to dBi. To convert a value in dBd to the equivalent dBi value, the following equation is utilized:

$$dBi = 2.14 + dBd$$

Table 2-5 can be used to help reinforce the conversion process that shows the calculated values along with the nearest approximate value found.

ERP and EIRP

Effective Radiated Power (ERP) and *Effective Isotropic Radiated Power* (EIRP) are the two most common references used for determining the transmit power of a communication site. ERP and EIRP are directly related to each other and a simple conversion can be achieved when one is known and the other is sought:

$$ERP = EIRP - 2.14$$

Table 2-6 can be used for a general comparison for various dB references. Although it is not complete, it covers most, if not all, the dB references that are encountered in the wireless industry.

Bandwidth on Demand and DBA

Bandwidth on demand is a phrase used to describe many different technologies that maximize the spectrum usage for an operator. Bandwidth on demand is more commonly associated with LMDS services, and provides a means for a subscriber to increase the amount of bandwidth they use only when needed. When the additional bandwidth is not needed, the excess capacity or bandwidth is made available for other subscribers to utilize.

Bandwidth on demand is also referred to as *Dynamic Bandwidth Allocation* (DBA). Many of the LMDS systems being deployed reference DBA as a feature or a future feature to the system. DBA or bandwidth on demand is an important aspect for oversubscribing the services (oversub-

Table 2-6	**dB**	**Reference**	**Comment**
dB reference table	dB	none	
	dBm	1mW	Standard wireless value
	dBs	1mW	Japanese wireless system reference
	dBc	none	Referenced to the carrier power
	dBw	1W	1 watt reference
	dBk	1KW	1 kilowatt reference
	dBu	1microvolt	Standard wireless value
	dBv	1V	1 volt reference

scribing is when the spectrum or bandwidth available can be allocated to many subscribers on the premise that they will not utilize the bandwidth at the same time).

For LMDS operators that have infrastructure designed to support only IP traffic where packets are sent, not a nailed-up connection, DBA can be said to exist for that system. However, many LMDS manufacturers do not support DBA at this time; they offer leased line services, in which activity factors on the time slots can be monitored and the bandwidth freed for other users on the network that could be sending voice or data.

OSI Levels

The OSI layers consist of seven layers, 1 through 7. OSI layer 7 is the highest and layer 1 is the lowest and is the physical layer that is the starting point for the OSI layer. See Table 2-7.

LMDS systems for the wireless access portion utilize layer 2 and layer 3 of the OSI architecture. From the concentration node's point of view, all OSI levels could be and are normally utilized.

Table 2-7

OSI layers

OSI Layer	Layer Name	Comments
7	Application	Used for connecting the application program or file to a communications protocol
6	Presentation	Performs the encoding and decoding functions
5	Session	Establishes and maintains the connection for the communication processes in the lower OSI layers
4	Transport	Error correction and transport, both Tx and Rx are performed here
3	Network	Switching and routing functions for the MSC are done here
2	Data Link	Receives and sends data over the physical layer
1	Physical	Actual media used for sending and receiving the communications (radio or fiber optic wires are two examples)

Pulse Code Modulation (PCM)

PCM is a process for digitizing and encoding analog signals and is used, along with other processes, in T-carrier-based transmission systems. T-carrier-type systems are the most widely used for interconnecting switches. Figure 2-6 shows a functional block diagram of the PCM process.

The PCM Process Step One—Sampling

The first step in the PCM process is called sampling, which involves choosing and measuring points along the analog speech signal. The actual measured values are called samples. The samples produce a series of pulses that represent the amplitudes of the various signals at specified known times. In order for this process to take place, the rate at which to sample (time intervals between successive measurements) the original analog voice signal needs to be determined. This rate is dependent upon the Nyquist theory and upon the voice grade bandwidth, as determined by the telecommunications industry. If the sampling rate is great enough (two times the highest frequency of the signal to be sampled), it is possible for the receiving equipment of a communications system to reconstruct the original ana-

Figure 2-6
PCM functional
diagram

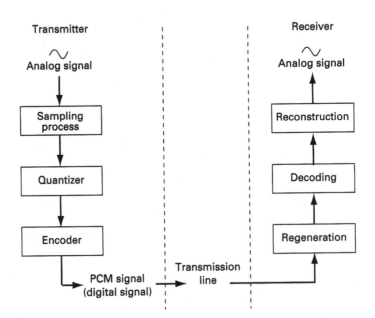

log signal based upon the sampled signal data. Although the industry bandwidth limits for voice grade analog signals can range from 300Hz to 3,000Hz, the actual upper limit is set at 4,000Hz. Given this rule and data, the sampling rate is therefore, twice the bandwidth of the upper limit (4,000Hz times 2 equals 8,000 samples per second).

After the sampling process is conducted upon the analog signal, as shown in Figure 2-7, it yields a digital signal that represents only one-half the original signal. This digital signal will not resemble the voice frequency signal; instead, it will appear as a series of pulses whose height will be the same as the original analog signal at the points where the samples were measured.

The analog signal that has been divided into a series of short sampled pulses is referred to as having been *Pulse Amplitude Modulated* (PAM).

The PCM Process Step Two—Quantizing

The next step is the quantization process. This process takes the amplitudes of the samples generated in step one and quantizes them so that they

Figure 2-7
PCM sampling
process

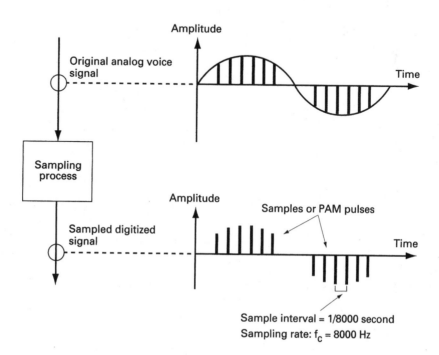

Amplitude

Original analog voice signal

Time

Sampling process

Sampled digitized signal

Amplitude

Samples or PAM pulses

Time

Sample interval = 1/8000 second
Sampling rate: f_c = 8000 Hz

can be represented by an eight-digit binary digital code. The quantization step, or quantizer, assigns the amplitude ranges of the PAM signal into a finite number of amplitude values. This is completed by dividing the total amplitude range into intervals. Numerical values are assigned to each interval, depending upon its distance (positive or negative) from the time axis. These values form quantizing intervals for the PAM samples to fall within. Each sample will be assigned the value of a quantizing interval, depending upon the one it is closest to in height. Because the heights of the PAM samples will rarely match the heights of the quantizing intervals exactly, there will be a certain amount of error or rounding of the samples. This error (quantizing error) is irretrievable and adds noise to the signal (quantizing noise). This noise is heard as a hissing noise in an actual telephone. See Figure 2-8.

The m-255 or m-Law is the Bell standard quantization process for use in the United States. This specification allows for better resolution of the lower quantization levels that are more predominant in PAM voice signals. The European standard is the A-Law, which provides a slightly better signal-to-noise ratio for small signals, but has a higher idle channel noise level.

Figure 2-8
PCM quantizing process

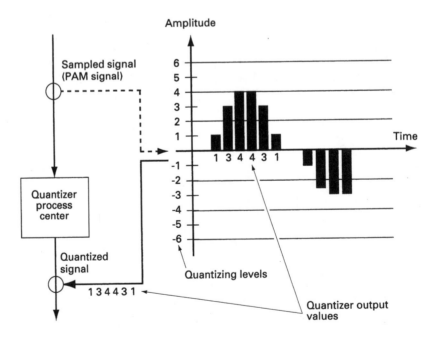

The PCM Process Step Three—Encoding

The final step of the PCM process, the encoding step, takes the output values of the quantizer and converts them into a binary serial data stream that is capable of being transmitted over a single transmission line as shown in Figure 2-9.

This process is then reversed at the receiving end of the communication system to yield the original analog signal.

T1

A T-Carrier is also called a T1, a DS1, or even a span. The T-1 carrier is the digital, two-way transmission of voice, data, or video signals over a single

Figure 2-9
PCM encoding
process

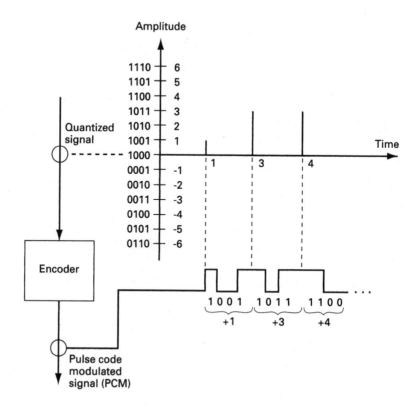

high-speed circuit. The actual data transmission rate of the T-1 carrier system is based upon the bandwidth utilized to transmit one digitized voice signal. This signal is referred to as DS-0 (digital signal, level zero). The bandwidth of this DS-0 signal is equal to 64,000 bits per second. The initial T-1-based systems transmitted 24 voice channels over two pair of twisted cables at a rate of 1,536,000 bits per second. Later systems introduced a necessary control bit for synchronizing T-1 multiplexing equipment. This additional bandwidth brought the T-1 carrier transmission rate up to 1,544,000 bits per second (also denoted as 1.544 Mb/s). This is now known as DS-1, the first signal level of a T-1 carrier system. Table 2-8 presents the various signal levels and transmission rates for T-carrier-based systems.

E1

The E1 carrier is the digital, two-way transmission of voice, data, or video signals over a single high-speed circuit whose format is found throughout the world, except in North America. The actual data transmission rate of the E1 carrier system is based upon the bandwidth utilized to transmit one digitized voice signal. This signal is referred to as DS-0 (digital signal, level zero). The bandwidth of this DS-0 signal is equal to 64,000 bits per second.

The initial E1-based-systems transmitted 30 voice channels plus two signaling channels over two pair of twisted cables at a rate of 2.048 M bits per second, which is typically called a 2Mbps line. The E1 signaling hierarchy is very similar to the T1 system utilized in North America, except for the bandwidth.

Table 2-8

T carrier

Signal Level	Carrier System	Number of T-1	Voice Circuits	Mb/s
DS-1	T-1	1	24	1.544
DS-1C	T-1C	2	48	3.152
DS-2	T-2	4	96	6.312
DS-3	T-3	28	672	44.736
DS-4	T-4	168	4,032	274.760

Table 2-9

E carrier

Carrier System	Number of E-1	Voice Circuits	Mb/s
E-1	1	30	2.048
E-3	16	480/496 BRI (PRI)	34.368

The Table 2-9 presents the various signal levels and transmission rates for E-carrier-based systems.

If the circuit is conditioned as a PRI for the E1, then a total of 30 BRIs plus 1 DSO for the D channel are used (30B +D). Obviously, not all of the channels have to allocated for the PRI.

Description of the SS7/CC7 Data Network

Signaling System 7 (SS7) and *Common Channel Signaling* (CC7) are popular data communications transfer protocols used to provide out-of-band signaling for processing calls to the PSTN and CLEC networks. Depending on the type of services the LMDS system will make available to the subscriber base, the use of SS7/CC7 signaling might be required. For instance, the use of the SS7/CC7 network is used for ISDN-related call functions, advanced call features, and for regular voice communications, which are still the the predominant communication media.

The SS7/CC7 protocol has three major advantages over in-band signaling. The first advantage is the improved post dial delay for faster call setup times. This network improvement is very noticeable to the end user. The second advantage is the capability to signal in the reverse direction, improving communication between various nodes in the network. The third advantage of SS7/CC7 over conventional in-band signaling is the improvement in the network fraud prevention due to the data (voice) messages and signaling messages being sent over separate routes.

There are three main components of an SS7/CC7 network:

- Signaling Service Points (*SSPs) and* Signaling Points (*SPs*) SSPs and SPs serve as the connecting points for end subscriber units, such as land line and mobile phones. These nodes perform call processing on calls that originate, terminate, or tandem at the location.

- Signal Transfer Points (*STPs*) STPs transfer SS7 messages between interconnected nodes based upon information contained in the SS7 address fields.

- Signaling Control Points (*SCPs*) SCPs operate in a similar manner as the SSPs and SPs, but only process database information functions.

The SSPs, SPs, STPs, and SCPs are connected by links, usually link pairs, which are used to transport information to and from each of the elements. There are several types of links that are used in a SS7/CC7 network, which are listed as follows. Their general functionality and location within a SS7/CC7 network are also briefly covered.

- *Access Links ("A" links)* Access links carry SS7 messages between SSPs and STPs, and between STPs and SCPs.

- *Bridge Links ("B" links)* Bridge links carry SS7 messages between STPs in different regions of the same network.

- *Control Links ("C" links)* Control links carry SS7 messages between mated STPs.

- *Diagonal or Quad Links ("D" links)* Diagonal or quad links carry SS7 messages between STPs in networks.

- *Extended Links ("E" links)* Extended links carry SS7 messages between SSPs and remote STPs.

- *"F" Links ("F" links)* "F" links carry SS7 messages between SSPs.

The links are referred to as a linkset, which consists of two or more links connecting adjacent nodes and sharing the same routing information.

Figure 2-10 shows the use of these components in an SS7 network.

There are a number of protocols that can be supported by the SS7/CC7 network. Although the SS7/CC7 network protocol provides the reliable transport of data messages between network nodes, there are additional protocols that can be used to build and decode these various messages, based upon the particular application in the network. The use and function of these additional protocols of an SS7/CC7 network follows the *Open System* (OSI) Interface model shown in Figure 2-11. This model consists of seven layers, described as follows:

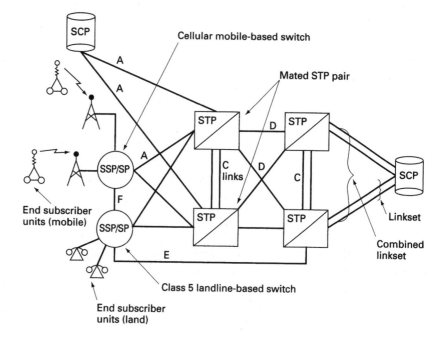

Figure 2-10
Example of a basic
SS7 network

- *Physical layer* This layer provides the electrical and mechanical interface to the network transmission facilities. This layer defines the protocol for actually transmitting a stream of serial data bits between two network nodes.

- *Data link layer* This layer provides control for the transmission, framing, and error-correction functions over a single transmission data link facility.

- *Network layer* This layer provides the functions to establish clear, logical, and physical connections (if required) across the network. Included in these functions are network routing (addressing) and flow control functions across the computer to network type interfaces.

- *Transport layer* This layer provides independent and reliable message inter-exchange functions to the upper three application-orientated layers. This layer basically acts as an interface between the bottom three layers and the upper three layers.

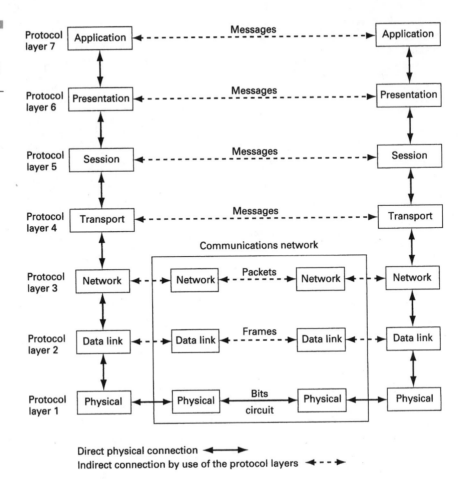

Figure 2-11
Open System
Interconnection (OSI)
model

- *Session layer*   This layer provides the functions that control the data exchange process. It basically sets up and clears communication channels between two communicating network entities.

- *Presentation layer*   This layer provides the syntax translation between two applications communicating over the network.

- *Application layer*   This layer provides the user interface to a variety of network services and also manages the communication between applications on two communicating systems. Example applications include *file transfer access and management* (FTAM) applications and electronic mail.

Figure 2-12
SS7 protocol structure

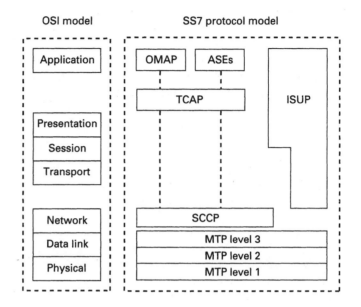

The six layers pertaining to the SS7/CC7 network are listed and described as follows and in Figure 2-12:

- The *Message Transfer Part* (MTP) is the lowest and most basic layer in the network protocols. All other parts of the SS7 network use MTP for basic transport between SPs. MTP is comprised of the physical (data link) layer and the network layers of the OSI model.

- The *Signaling Connection and Control Part* (SCCP) corresponds to the transport layer of the OSI model. It provides support for both connectionless and connection-oriented network services.

- The *Telephony User Part* (TUP). This part provides for call setup and teardown under the SS7 network protocol.

- The *ISDN User Part* (ISUP) provides interoffice ISDN services and uses either SCCP or MTP for message transport.

- The *Transactions Capabilities Part* (TCAP) uses SCCP for message transport. TCAP supports the transfer of messages not associated with circuit control.

- The *Operations Maintenance and Administration Part* (OMAP) provides the application protocols to monitor and coordinate all the network resources.

Time Division Multiplexing (TDM)

In the TDM process, the large-capacity transmission channel is divided up into time intervals called time slots instead of the frequency slots mentioned in the FDM process. Next, PAM samples from the PCM process are interjected into a single high-capacity (and higher-speed) transmission channel. This process of interjecting PAM samples (pulses) of each individual input channel into the high-speed transmission channel occurs once every sample cycle. The total PAM samples or pulse signals embedded into the high-speed channel during one cycle is called a frame. On the receiving end of the transmission facility, this process is reversed (demultiplexed) and the original signals are reconstructed.

The final output of the FDM process is a TDM-PAM signal. Taking this concept further, if the FDM process were conducted by not using the PAM samples, but with the PAM encoded word representing the PAM sample, then the resulting final output would be a TDM-PAM signal. This subprocess is called time slot interleaving. Most PCM systems used in the telecommunications industry are of the TDM-PAM system type. See Figure 2-13.

VoIP

Voice over IP (VoIP) continues to provide a viable alternative for call delivery for voice traffic. It is interesting that most of the initial VoIP implementations have not occurred over the Internet but rather over corporate LANS and private IP networks such as *Long Distance* (LD) providers This has mitigated the QoS problems associated with VoIP on the Internet.

When mentioned in many circles, VoIP invokes quality concerns due to delay and jitter problems, thus QoS, when the access media is over the public Internet. As mentioned previously, although the true application for VoIP is a transport medium over private or dedicated pipes or networks, where the QoS issue no longer is an issue.

The original standards activity for VoIp was defined in H.323, "Packet-Based Multimedia Communication Systems." This standard's wide use was a direct result of Microsoft offering it as freeware. There is, however, an alternative standard that is currently in competition with H.323: *Media Gateway Control Protocol* (MGCP)—also called *Single Gateway Control Protocol* (SGCP).

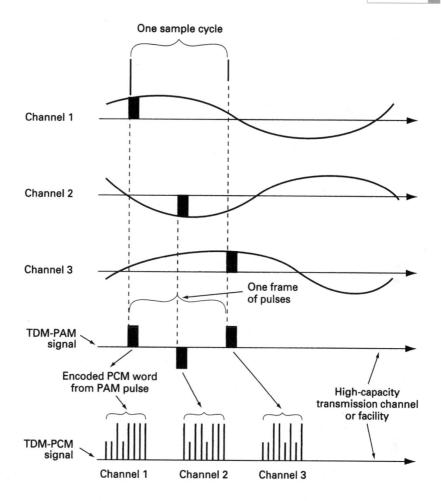

Figure 2-13
Time Division
Multiplexing (TDM)
example

SGCP assumes a control architecture that has a similar architecture to the current PTT voice system, where the control is done outside the gateway itself. The external call control elements are referred to as call agents.

LMDS and CLEC operators that utilize infrastructure that is IP-based only can also offer voice services as part of their offering if the proper QoS and delivery issues are addressed in the design and service offering. For the LMDS operators that offer voice via CES, voice services are an attractive entry point for customers. However, the discussion of VoIP does not need to be conveyed to the customer if the proper delivery and QoS issues are addressed.

A primary reason why VoIP is so attractive for an LMDS operator is not solely related to the interconnect savings that may be achieved, but in saving the spectrum because IP traffic is by itself dynamic bandwidth.

Figure 2-14 is a depiction of the major components involved with providing VoIP, either as a direct service or as an alternative transport medium that the LMDS operator uses to be more cost-competitive or better yet improve the margin.

In Figure 2-14, VoIP can be delivered either directly to a public data network or the Internet, depending on the SLA that is used. In addition, the figure depicts the issue of the operator using VoIP as a medium for handling voice traffic into the switching complex, where it then converts the IP traffic into classical TDM traffic for interfacing to the PTT for call delivery.

Figure 2-15 depicts the connection between an LMDS operator in one market and its operation in another market. The diagram can of course be meant for an ISP, CLEC, or large corporation.

VOIP is the most flexible choice for voice transport because it can run over any layer 1 or layer 2 infrastructure. This flexibility is particularly important in heterogeneous environments such as LMDS systems.

Cable Systems

Cable operators have a unique advantage over LMDS operators regarding the residential market because they have presence in many residential homes. The issues may, however, be the quality of the cable plant itself, which dictates the delivery of services that can effectively be offered. The issues with quality of the cable plant are primarily driven by the amount of

Figure 2-14
VoIP network

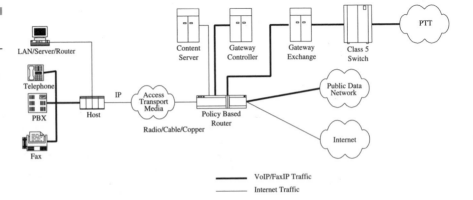

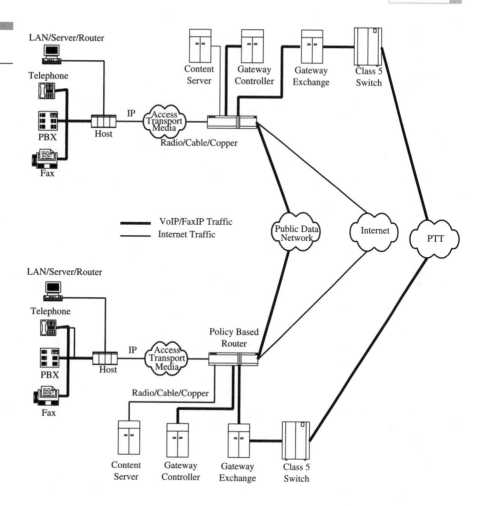

Figure 2-15
VoIP network

drops that are on any cable leg that directly impacts the ingress noise problem, limiting the capability for the cable plant to provide high-speed, two-way communication. Because most of the information flow is from the head end to the subscriber, the system does not have to support symmetrical bandwidth requirements.

HFC

A *Hybrid Fiber/Coax* (HFC) network is shown in Figure 2-16, with the enhancement of providing two-way communication for both voice and data, besides the video service offering. The primary access method is physical

Figure 2-16
HFC

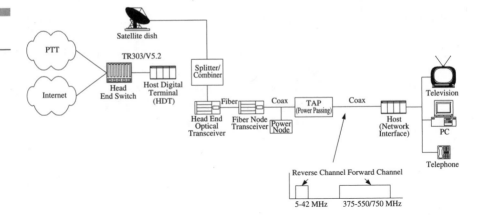

Figure 2-16
HFC

Figure 2-17
RAD/RASP

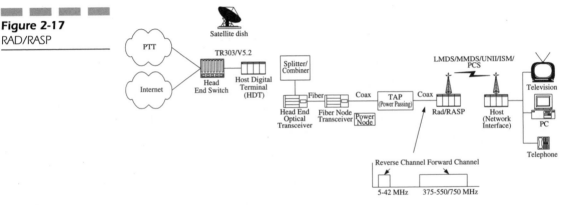

media where the connection made to the subscriber at the end of the line is via coaxial cable. For increased distance and performance enhancements, fiber optic cables are often part of the cable network's topology.

Figure 2-17 is an example of a cable operator utilizing wireless access as the last leg in the access system. The wireless device listed can be a base station or a small RAD/RASP unit installed on the coaxial cable itself. The figure depicts the potential for a cable operator and a LMDS operator to utilize each other's infrastructure in order to deliver services.

PCS is listed in the figure not as a provider for mobility telephony, but as a better allocation for the PCS C-band that was auctioned in the U.S. for delivering last mile bandwidth services.

FDD Point to Multipoint

Frequency Division Duplex (FDD) utilizes two separate radio channels for communicating between the base station and the host terminal. One of the radio channels, f1, is for the communication link, downlink, from the base station to the host terminal. The other radio channel, f2, is for the communication link, uplink, from the host terminal to the base station. See Figure 2-18.

The channels f1 and f2 are normally spaced a distance apart for isolation purposes. The FDD system utilizes dedicated channels for uplink and downlink communication. Figure 2-19 illustrates how the uplink and downlink channels are paired.

Many of the deployed LMDS systems utilize a FDD system. The particular access scheme that is used with FDD systems is quite varied. Some of the variants involve a TDMA channel as the downlink and a FDMA channel on the uplink. Another popular access method is to utilize TDMA on the downlink and ATM on the uplink.

Figure 2-18
FDD system

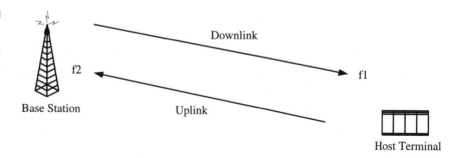

Figure 2-19
FDD spectrum
allocation example

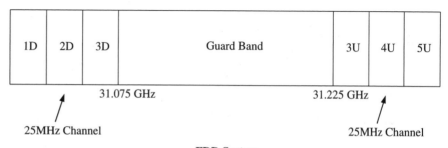

FDD System

FDD systems, no matter what access scheme is utilized, can be frequency-coordinated between operators utilizing the same frequency band, channels following industry-recognized methods and procedures. Where the trick comes in for frequency coordination is when an FDD and a TDD system must coexist in the same market or an adjacent market where frequency coordination is required. Specifically, the frequency coordination for the uplink on the FDD system is also a downlink channel for the TDD system.

TDD Point to Multipoint

Time Division Duplex (TDD) utilizes one radio channel for communication between the base station and the host terminal. The duplexing that is done is based on time, not frequency, as typically is done with a FDD system. The same channel f1 is used for both uplink and downlink communication between the base station and the host terminal as shown in Figure 2-20.

By its nature, TDD is designed to be more spectrally efficient than a FDD system when involving data communication that is non-symmetrical like Internet traffic. If the traffic is non-symmetrical, as IP Internet traffic is (where the downlink accounts for 75% of the traffic), then four channels are used for TDD, which is the same as two channels for FDD. See Figure 2-21.

The complexity with the TDD system versus a FDD system lies in the interference issues when the system is deployed in a mixed technology market—that is, TDD and FDD systems. The TDD system for both the base station and the host terminals needs to be coordinated with the FDD system because the host terminals will be transmitting when utilizing TDD on a FDD receive channel.

Figure 2-20
TDD system

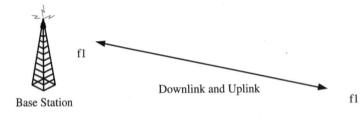

Base Station

f1

Downlink and Uplink

f1

Host Terminal

Figure 2-21
TDD spectrum
allocation example

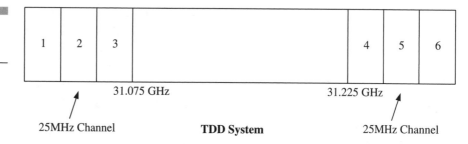

| 1 | 2 | 3 | | 4 | 5 | 6 |

31.075 GHz   31.225 GHz

25MHz Channel   **TDD System**   25MHz Channel

Figure 2-22
WLAN

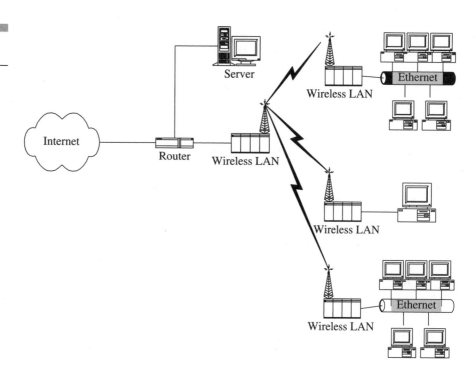

Wireless LAN

Wireless LAN (WLAN) is another wireless platform that enables various computers or separate LANS to be connected together into a LAN or a WAN. The big advantage is that WLAN-enabled devices do not need to be physically connected to any wired outlet, enabling flexibility for location, as shown in Figure 2-22.

There are several protocols that fall into the WLAN arena. The most prevalent is IEEE 802.11, but Bluetooth is also being referred to as a WLAN protocol. Both 802.11 and Bluetooth operate in the 2.4Ghz *industrial, scientific, and medical* (ISM) band.

802.11 was designed initially as a wireless extension for a corporate LAN for enterprise applications and has numerous devices that have been manufactured to this specification. Basically, the IEEE 802.11 protocol is a shared medium and utilizes CSMS/CA, which is a listen before talk protocol, with standards for collision sense multiple access/collision avoidance.

Table 2-10 is a simple comparison between the IEEE802.11 protocol and Bluetooth. Although both utilize the ISM band their format and purpose is different. The IEE802.11 devices are meant to cover a wider area than Bluetooth devices and the 802.11 devices have the potential of higher throughput. The data rate in the chart for IEEE802.11 shows a range in speeds that are dependent upon the modulation format used, power, and interference experienced.

References:

Brodsky, Ira. "3G Business Model." *Wireless Review*, June 15, 1999, 42.

Carmeli, Alon. "Gaining Momentum Using Two-Way Coax Plant for Data." *Communication Engineering and Design*, April 1997, 54.

Channing, I. "Full Speed GPRS from Motorola." *Mobile Communications International*, Number 65, October 1999, 6.

Channing, I. "It's WLL, Jim, but not as we know it." *Mobile Communications International*, Number 65, October 1999, 89.

Table 2-10

WLAN

WLAN	802.11	Bluetooth
Transport	2.4GHz ISM DSS	2.4 GHz ISM FHSS
Data Rate	1–11 Mbps	1Mbps
Range	50 m	1–10m
Connecting Devices	128	8
Power	+20 dBm	0 dBm

Daniels, Guy. "A Brief History of 3G." *Mobile Communications International*, Number 65, October 1999, 106.

Gull, Dennis. "Spread-Spectrum Fool's Gold?" *Wireless Review*, January 1, 1999, 37.

Hartman, Dave and Ted Rabenko. "Allocating Resources Over a DOCSIS Network." *Communications Engineering & Design*, December 1998, 76.

Hudec, Premysl. "Noise in CATV Networks." *Applied Microwave & Wireless*, September 1997, 22.

Iler, David. "Telephony over Cable." *Communications Engineering & Design*, December 1998, 22.

Jacobs, Jeffrey and Patrick Brown. "Life-Cycle Cost Savings in HFC Networks." *Fiberoptic Product News*, 8th edition, 1998, 55.

Kegerise, Craig, Steve Joiner and Joel Rosson, "The Next Generation of Fiber-Based Networks." *Fiberoptic Products News*, 8th edition, 1998.

Kesten, Gayle. "How to Construct an E-Commerce Site." *Varbusiness*, June 7, 1999, 79.

Masud, Sam. "Cable Operators Eye Telephony." *Telecommunications*, November 1999, 35.

Oba, Junichi. "W-CDMA Systems Provide Multimedia Opportunities." *Wireless System Design*, July 1998, 20.

O'Keefe, Sue. "TDD vs FDD: The next hurdle." *Telecommunications*, December 1999, 40.

Salter, Avril. "W-CDMA Trial & Error." *Wireless Review*, November 1, 1999, 58.

Shank, Keith. "A Time to Converge." *Wireless Review*, August 1, 1999, 26.

Smith, Clint and Curt Gervelis. *Cellular System Design and Optimization*. New York, NY: McGraw-Hill, 1996.

Smith, Clint. *Practical Cellular and PCS Design*. New York, NY: McGraw-Hill, 1997.

Stanton, Steve. "Testing W-CDMA Places New Demands on Designers." *Test/Measurement*, July 1998, 37.

Webb, William. "CDMA from WLL." *Mobile Communications International*, January 1999, 61.

Wesley, Clarence. "Wireless Gone Astray." *Telecommunications*, November 1999, 41.

Wirbel, Loring. "LMDS, MMDS Race for Low-Cost Implementation." *Electronic Engineering Times*, November 29, 1999, 87.

Witowsky, William. "VoP: Standards Remain Elusive." *Telecommunications*, May 1999, 53.

## CHAPTER 3

Radio Elements

Introduction

As mentioned previously, an LMDS system has many of the same fundamental building blocks that any wireless system has. However, the basic premise of LMDS is to provide the last mile or last kilometer of access through the use of radio. Because radio is a critical element in the successful deployment of an LMDS system, a brief discussion regarding the physical components of a radio system needs to be included here. An LMDS system utilizes radio frequency spectrum as the transport mechanism. The advantage that LMDS has over other many wireless access systems is that the end user is stationary and not mobile, thereby eliminating many of the complications that take place with wireless systems. However, the fundamental transport medium is still radio, so it is essential that the designer understands many of the fundamental concepts of radio engineering in order to select the best components for the system and optimize the network for maximum throughput and reliability.

The fundamental building blocks of a communication system are shown in Figure 3-1.

The simplified drawing in Figure 3-1 shows the major components in any communication system: antenna, filters, receivers, transmitter, modulation, demodulation, and propagation. Each of the major components identified in the figure requires the RF Engineer to consider them all in order to achieve the optimum design for the situation.

Entire books are written about each section listed in Figure 3-1. It is essential to know all the major components that actually make up the communication system being designed. For instance, if the design of the system involved using 7Mhz channels, it would make little sense if the system were designed using a 12.5MHz channelized system, even if both are used in LMDS systems. Therefore, knowing the design characteristics of each of the components is essential for building a communication system that will provide the proper transport functions for the information content.

Figure 3-1
Communication
system block diagram

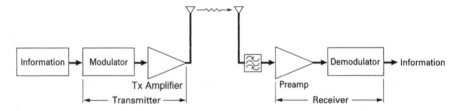

So, where do we begin the brief discussion of the major elements associated with the radio components of a LMDS system? To start, we will begin by assuming that there is some type of information that needs to be transported across the radio system, either to or from the base station. The information is not relevant for reviewing the major components of the radio system because the radio system's job is to transport the information within the given bandwidth and at the correct frequency band. What is important to realize is that the choice of modulation formats available need to be reviewed, based on the bandwidth available with a desired throughput for information.

One key concept is that for any modulation format chosen, for a system to transport more information, it requires more bandwidth. However, the increase in bandwidth also decreases the receiver sensitivity, as well as requiring more power from the transmitter. Infinite bandwidth is not available in all cases, so the modulation format needs to be increased to increase the throughput for data with a constant bandwidth. But with increased modulation formats (for example, going from 4QAM to 64QAM), both range and power may be sacrificed to maintain the same reliability factors for the system.

Therefore, it is essential for the design engineer to understand some of the basic building blocks of a radio system so that proper decisions can be made. The discussion will begin with the transmitter, progress to general modulation types, discuss antennas, and finally progress to the receiver.

Transmitters

There are many types of transmitters or amplifiers, in a communication system. Most of the amplifiers are located in the receiver for the communication cell site. The fundamental difference between the amplifiers in the transmit portion and the receiver lie in the power type they are able to amplify and deliver to the desired load. The focus of this section will be toward RF transmitters, which are power-oriented.

Knowledge of RF transmitter types and different combining techniques will enable an RF engineer to maximize the efficiency of a communication site through improving the amount of energy delivered to the antenna system or reducing the amount of physical antennas at a site. The amount of physical antennas available at a site may or may not be driven by economic reasons alone; local ordinances may have a more profound role in deciding the amount of ultimate configuration for a communication site.

Transmitter System Building Blocks

The transmitter block diagrams for the three basic forms of radio communication are included next for reference. Figure 3-2 is a block diagram of an AM transmitter. The AM transmitter changes the amplitude of the carrier as a function of the information content.

Figure 3-3 is a brief block diagram of an FM transmitter. The FM transmitter modulates the information content by changing the frequency of the carrier as a function of the information content.

Figure 3-2
AM modulation

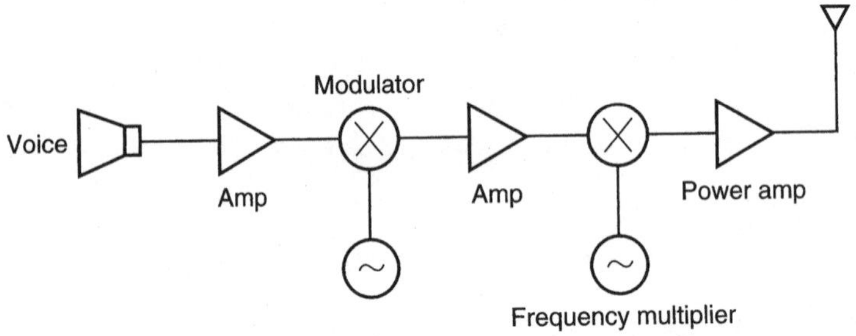

Figure 3-3
FM modulation

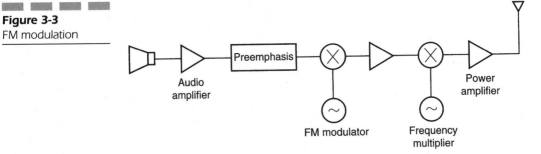

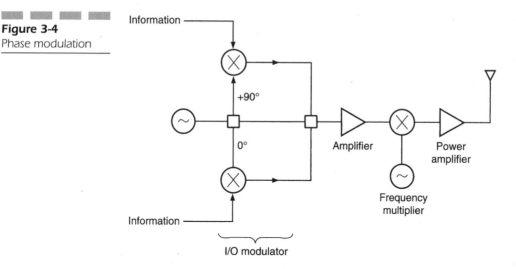

Figure 3-4
Phase modulation

Figure 3-4 is a block diagram of a PM transmitter. The PM transmitter places the information onto the carrier, as it is done with FM. However, the modulation is achieved through adjusting the phase of the information that rides on the carrier.

Modulation

To convey data and voice information from one location to another without physically connecting them together as in LMDS, it is necessary to send the information by another method. There are many methods for conveying information to and from locations that are not physically connected. Some of the methods include talking, flags, drums, and lights. Although each of these methods has its advantages and disadvantages, the common communication problem they have involves the physical distance the sender and the receiver have to be from each other, along with the information throughput.

In order to increase the distance and increase the information transfer rate between sender and receiver, the use of an electromagnetic wave is employed for this application. The use of electromagnetic waves is fundamental to radio communication, but it is necessary to modulate the carrier wave at the transmitting source and then demodulate it at the receiver. The

Figure 3-5
Basic radio system

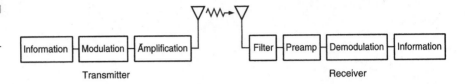

Figure 3-6
Modulation
techniques

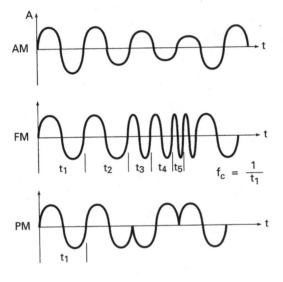

modulation and demodulation of the carrier wave forms the principle of a radio communication system.

The generalized radio system is shown in Figure 3-5.

The choice of modulation and demodulation utilized for the radio communication system is directly dependent upon the information content to be sent, the available spectrum to convey the information, and the cost. The fundamental goal of modulating any signal is to obtain the maximum spectrum efficiency, or information density per hertz.

Although there are many types of modulation and demodulation formats utilized for the conveyance of information, all the communication formats rely on one, two, or all three of the fundamental modulation types. The fundamental modulation techniques are *Amplitude Modulation* (AM), *Frequency Modulation* (FM), and *Phase Modulation* (PM). Figure 3-6 highlights the differences between the modulation techniques in terms of their impact on the electromagnetic wave itself.

$$E(t) = A \sin (2\pi f_c t + \phi)$$

where A = amplitude
f_c = carrier frequency
ϕ = phase
t = time
E = instantaneous electric field strength

amplitude modulation (AM) modifies A
frequency modulatin (FM) modifies f_c
phase modulation (PM) modifies ϕ

Amplitude Modulation (AM) has many unique qualities, but it is not utilized directly in LMDS communication systems because it is more susceptible to noise. However, a variant of AM, QAM, is used predominantly.

Frequency Modulation (FM) is utilized for many mobile communication systems that employ analog communication. One common use of FM modulation is in an analog cellular communication because it is more robust to interference.

Phase Modulation (PM) is used for conveying digital information. There are many variations of phase modulation. Specifically, many digital modulation techniques rely on modifying the RF carriers' phase and amplitude, as in the generation of QPSK and QAM signaling formats.

There are a multitude of industry textbooks and technical articles that focus purely on each of the modulation schemes if more in-depth analysis is required.

Information Bandwidth

The information and the channels bandwidth is a critical parameter in determining which modulation scheme to utilize for the communication system. Often, it is the channel bandwidth that is defined and then the appropriate modulation technique must be applied. Conversely, the manufacturer of an LMDS system states the information throughput of their system for a given channel bandwidth.

The channel's theoretical capacity is defined by the Shannon-Hartley equation, shown as follows:

$$C = B \times \log_2 (1 + S/N)$$

$$C = \text{Capacity}$$

$$S/N = \text{signal to noise ratio}$$

$$B = \text{bandwidth}$$

Antennas

The antenna system for any radio communication platform utilized is one of the most critical and least-understood parts of the system. The antenna system is the interface between the radio system and the external environment. The antenna system for an LMDS system can consist of a single antenna or multiple antennas at the base station and one at the host terminal station. Primarily, the antenna is used by the base station site and host terminal for establishing and maintaining the communication link.

There are many types of antennas available, all of which perform specific functions, depending on the application at hand. The type of antenna used in an LMDS system operator can be a panel or horn, to mention a few. Coupled with the type of antenna is the notion of an active or passive antenna. The active antenna usually has some level of electronics associated with it to enhance its performance. The passive antenna is more of the classical type, in which no electronics are associated with its use and it consists entirely of passive elements. Most LMDS systems utilize an active antenna system because the Tx and Rx elements that are integrated into the antenna system on top of the mast itself.

There is also the relative pattern of the antenna, indicating in what direction the energy emitted or received from it will be directed. There are two primary classifications of antennas associated with directivity for a system: omni and directional. The omni antennas are used when the desire is to obtain a 360-degree radiation pattern; the directional antennas are used when a more refined pattern is desired. The directional pattern is what is used for an LMDS system, with the variants being associated with the physical gain, its apature, its vertical and horizontal beam widths, and of course its polarization.

The choice of which antenna to use will directly impact the performance of the base station, host terminal, or overall network. The radio engineer is primarily concerned with the design phase of the base and host stations because they are fixed locations—there is some degree of control over the performance criteria that the engineer can exert on the base station and potentially on the host terminal location.

The correct antenna for the design can overcome coverage problems or other issues that are trying to be prevented or resolved. The antenna chosen for the application must take into account a multitude of issues in the design phase, including the antenna's gain, its antenna pattern, the interface or matching to the transmitter, the receiver utilized for the site, the bandwidth and frequency range over which the signals desired to be sent will be applicable, its power-handling capabilities, and its IMD performance. Ultimately, the antenna you use for a network needs to match the system design objectives.

Antenna Performance Criteria

The performance of an antenna is not restricted to its gain characteristics and physical attributes; that is, maintenance. There are many parameters that must be taken into account when looking at an antenna's performance. The parameters that define the performance of an antenna can be referred to as the *figures of merit* (FOM) that apply to any antenna that is selected to use in a communication system.

There are many parameters and FOM that characterize the performance of an antenna system. The following is a partial list of the FOM for an antenna that should be quantified by the manufacturer of the antennas you are using. The trade-offs that need to be made with an antenna chosen involve all the FOM issues discussed in the following list.

■ *Antenna Pattern*   The graphical representation of the elevation and azimuth antenna patterns for both a omni directional and directional antenna is shown in Figure 3-7 through Figure 3-10. The antenna pattern is normally given in the form that is shown in. However, it is very important to note the reference scale that it is used because different scales can lead to different conclusions about how the antenna's pattern really will perform in the system.

The antenna pattern chosen should match the coverage requirements for the base station. For example, if the desire is to utilize a directional

Figure 3-7
Elevation antenna
pattern (omni
directional)

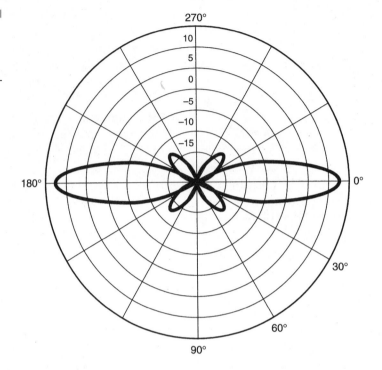

antenna for a particular sector of a cell site, say 30 degrees, choosing an antenna pattern that covers 90 degrees in azimuth would be incorrect. Care must also be taken in looking for electrical downtilt that may or may not be referenced in the literature.

- *Main Lobe* As shown in Figure 3-11, this is the radiation lobe containing the direction of maximum radiated power. The main lobe is referenced to the polarization for the antenna and simply reflects the directivity of the antenna. In this case, the polarization is vertical and therefore the main lobe representation is the elevation antenna pattern.

- *Side Lobe* As shown in Figure 3-12, this is the radiation's lobe in any direction other than the main lobe. The side lobes are important to consider because they create potential problems with generating interference. Ideally, there would be no side lobes for the antenna pattern. For downtilting, the side lobes are important to note because they can create secondary sources of interference.

- *Input Impedance* This is the impedance presented by the antenna at its terminals and it is usually complex. The input impedance for an

Figure 3-8
Azimuth antenna
pattern (omni
directional)

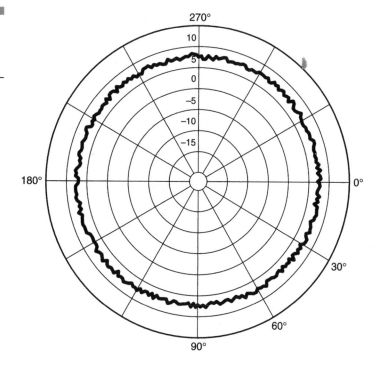

antenna should maximize the power transfer from the transmitter to the antenna for radiation. Nominal antenna input impedance is 50 ohms, which matches that of the coaxial cable used to connect the transmitter to the antenna. A mismatch in input impedance obviously impacts the energy transfer and decreases its overall power output.

- *Radiation Efficiency* This is the ratio of total power radiated by an antenna to the net power accepted by an antenna from the transmitter. The equation is as follows:

$$e = \text{Power Radiated}/(\text{Power Radiated} + \text{Power Lost}).$$

The antenna would be 100% efficient if the power lost in the antenna were 0. This number indicates how much energy is lost in the antenna itself, assuming an ideal match with the feedline and the input impedance. Using the previous equation, if the antenna absorbed 50% of the available power, it would only have 50% of the power for radiating and thus the effective gain of the antenna would be reduced.

Figure 3-9
Elevation antenna
pattern (directional)

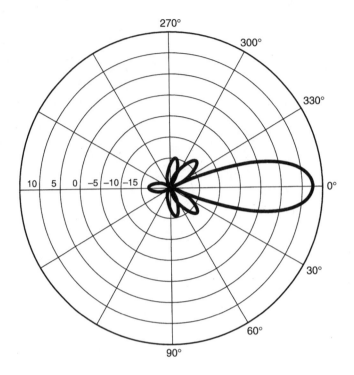

- *Beamwidth* This is the angular separation between two directions in which radiation interest is identical. The 1/2 power point for the beamwidth is usually the angular separation where there is a 3 dB reduction off the main lobe (see Figure 3-13 and Figure 3-14).

  Normally, the wider the beamwidth, the lower the gain of the antenna. A simple rule of thumb is for every doubling of the amount of elements associated with an antenna, a gain of 3 dB is realized. However, this gain comes at the expense of beamwidth. The beamwidth reduction for a 3 dB increase in gain is about one-half the initial beamwidth. Thus, if an antenna has a 12-degree beamwidth and has an increase in gain of 3 dB, its beamwidth now is 6 degrees.

- *Directivity* This is the ratio of radiation intensity in a given direction to that of the radiation intensity averaged over all the other directions. The equation for antenna directivity is as follows:

$$G(D) = \text{Maximum Power Radiation Intensity/}$$
$$\text{Average Radiation Intensity.}$$

Figure 3-10
Azimuth antenna
pattern (directional)

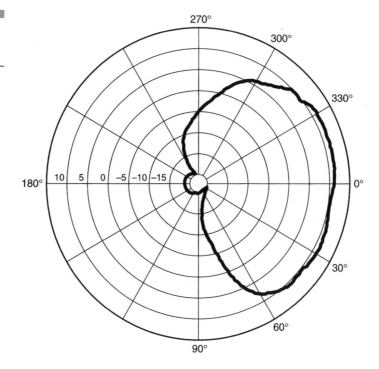

The directivity of an antenna can be improved though the use of
reflectors or its physical location to a leg of a tower. Often, the
directivity of the antenna pattern can be altered through the use
of the tower structure itself.

- *Gain*   This is a very important figure of merit. The gain is the ratio of
  the radiation intensity in a given direction to that of an isotropically
  radiated signal. The equation for antenna gain is as follows: G equals
  Maximum radiation intensity from antennas/maximum radiation from
  an isotopic antenna. The gain of the antenna can also be described as
  follows:

$$G = e \times G(D)$$

If the antenna were without loss, $e=1$, $G=G(D)$.

The gain for the antenna in the elevation plan is about 12 dB. It is
unclear, however, whether this gain is referenced to a dipole or isotropic
antenna.

Figure 3-11
Main lobe antenna
pattern

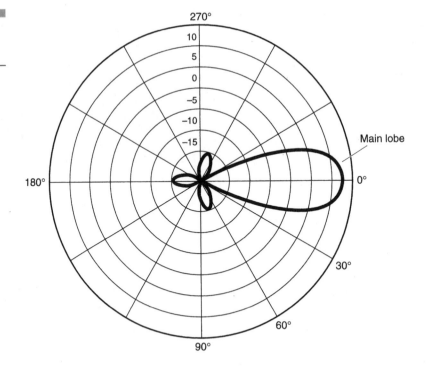

- *Antenna Polarization* The antenna polarization is defined as the polarization fields radiated by the antenna. The antenna's polarization is defined by the E field vector. Cellular and PCS systems utilize vertical polarization, whereas LMDS systems utilize vertical and horizontal polarizations.

- *Bandwidth* The bandwidth defines the operating range of the frequencies for the antenna. The SWR is usually how this is represented; besides the frequencies range, it is constant. A typical bandwidth that is referenced is the 1:1.5 SWR for the band of interest. Antennas are now being manufactured that exceed this (having a SWR value of 1:1.2 at the band edges).

  The antenna's bandwidth must be selected with extreme care to not only account for current but also future configuration options with the same cell site.

- *Front to Back Ratio* This is a ratio that deals with how much energy is directed in the exact opposite direction of the main lobe of the antenna. The front to back ratio is a loosely defined term. The IEEE Std 145-1983 references the front to back ratio as the ratio of

Figure 3-12
Side lobe antenna
pattern

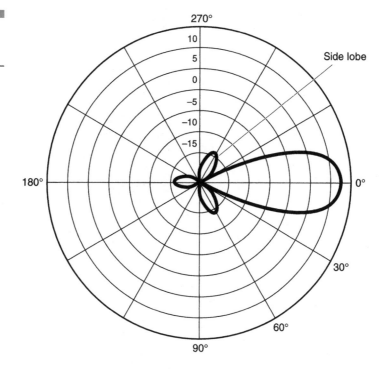

maximum directivity of an antenna to its directivity in a specified rearward direction. A front to back ratio is applicable only to a directional antenna because obviously there is no rearward direction with an omni directional antenna.

Many manufacturers reference high front to back ratios, but care must be taken in knowing just how the number was computed. For example, if installation is on a building and the antenna will be mounted on a wall, then the front to back ratio is not as important a FOM. However, if the antenna is mounted so there are no obstructions between it and the reusing cell, the front to back ratio can be an important FOM. In the latter case, the front to back ratio should be at least the C/I level required for operation in the system.

An example of the front to back ratio is shown in Figure 3-15.

The front to back ratio in Figure 3-15 is 27 dB.

■ *Power Dissipation* The total power that the antenna can accept at its input terminals is its power dissipation. This is important to note because the antenna chosen should be able to handle the maximum envisioned power load without damaging the antenna.

Figure 3-13
Elevation
beamwidth,
beamwidth = 20'

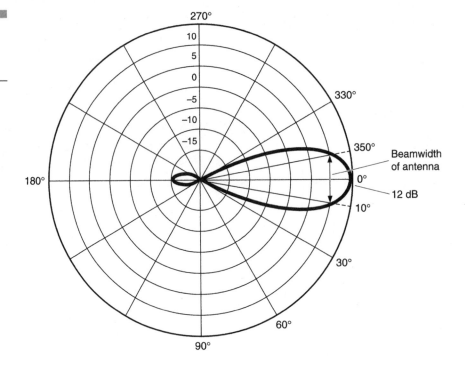

Figure 3-13
Elevation
beamwidth,
beamwidth = 20'

■ *Intermodulation*   Intermodulation by the antenna is introduced to the network in the presence of strong signals. The intermodulation that is referenced should be checked against how the test was run. For instance, some manufacturers reference the IMD to two tones, whereas some reference it to three or multiple tones. The point here is that the overall signal level that the IMD generates needs to be known, in addition to how many tones were used, their frequency of operation, the bandwidth, and of course the power levels that caused the IMD level.

■ *Construction*   The construction attributes associated with physical dimensions, mounting requirements, materials used, wind loading, connectors, and color constitute this FOM. For instance, some of the items that need to be factored into the construction FOM are the use of materials, whether the elements are soldered together or bolted, the type of metals that are used, and what their life expectancy is in the environment in which the antenna will be deployed. For instance, if you install antennas near the ocean or an aircon unit that uses salt water for cooling, then it is imperative that the material chosen will not corrode in the presence of salt water.

Figure 3-14
Azimuth beamwidth,
beamwidth = 75'

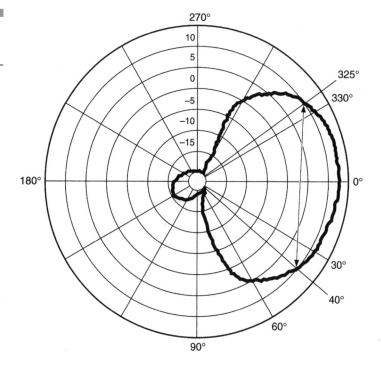

- *Cost* How much the antenna costs is a critical FOM. Now matter how well an antenna will perform in the system, the cost associated with the antenna will need to be factored into the decision. For example if the antenna meets or exceeds the design requirements for the system, but it costs twice as much as another antenna that meets the requirements, the choice seems obvious—pick the antenna that meets the requirement at the lowest cost.

Filters

Filters play an integral part in the design and operation of a radio and communication system. There are many types of filters that can be deployed. The selection of which filter to utilize is based on its mission statement and cost, either in terms of spectrum or actual monetary issues. Simply put, the purpose of a filter is to allow the desired energy or information to pass undistorted in phase, amplitude, or time, and at the same time completely suppress all other energy.

Figure 3-15
Front to back ratio

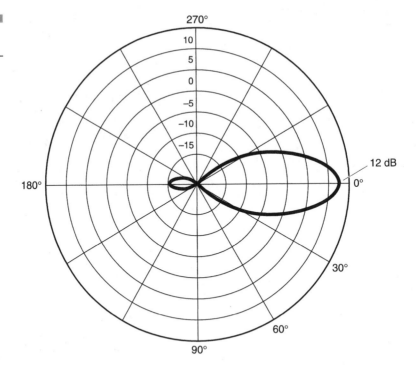

More often than not, the filter characteristics of an existing or new communication system are overlooked by the system design engineer. With the proliferation of wireless communication involved with LMDS and other sister bands, the need to pay particular attention to the filter characteristics of the base station and host terminals becomes more paramount. With this proliferation comes the demand for smaller physical size, with increased attenuation characteristics of unwanted signals while not distorting the desired signal in any fashion.

There are many types of filters available for use and each has its own unique characteristics. The specific characteristics of a filter are driven by its physical construction. The physical construction of the filter is an important aspect of the selection process and the types of filters listed are passive filters only. Although active filters have many good applications, they are not covered in this section.

There are four general classifications of filters that are used throughout all of radio communications. The following filter classifications are listed for reference. It should be noted that there are many perturbations regarding combinations of the general filter types listed. The specific configuration chosen is entirely dependent upon the application that it meant to solve and the acceptable trade-offs that come along with the filter choice made.

The general classification for filters falls into the following four basic types:

1. Low Pass Filter
2. High Pass Filter
3. Bandpass Filter
4. Band Reject Filter (Notch Filter)

Ideally, a filter would pass without attenuation to all the frequencies within a specified passband and infinitely attenuate all those frequencies outside of the passband. Additionally, the time response of the filter would be such that the output is identical to the input with some delay time. In other words, the transfer function for the filter should be equal to one for the frequencies of interest only.

An example of an ideal low pass, high pass, bandpass, and band reject filter is included in Figure 3-16.

As is often the case with the acquisition of a wireless system such as LMDS, the filters used have been predefined by the manufacturer of the equipment. However, it is important to understand the fundamental implications associated with filter techniques for both the base station and host terminal that could improve or degrade the system's performance. Because the ideal filter is not realizable at this particular time, the specific filter

Figure 3-16
Ideal filters (a) low pass (b) high pass (c) bandpass (d) band reject

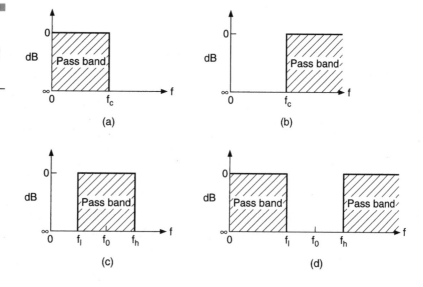

type selected and used in an LMDS system should be based on which trade-offs or imperfections can be best tolerated.

General Characteristics

The following is a general selection process for determining the type of general filter needed for the application. However, for certain applications, several filters can be cascaded together to solve a particular design issue.

The general rules are as follows:

Low Pass Filters:

High Frequency Interference Rejection

Band Limiting

Harmonic Suppression

High Pass Filters:

Band Limiting

Noise Reduction

Interference Elimination

Broadcast Signal Conditioning

Bandpass Filters:

Band Limiting

Comb Filter

Interference Elimination

Notch Filter:

Selective Frequency Rejection

Noise Reduction

Interference Elimination

Filter Performance Criteria

There are many aspects and criteria that define a filter's performance. The first and most important is the mission statement the filter itself is trying to resolve. Specifically, if the design calls for a filter to pass frequencies between 24.5GHz and 25.5GHz, a bandpass filter might be the best general type to utilize. However, the other attributes for the filter will determine just how well the filter performs in the application.

When selecting a filter, the following are some of the criteria that you should define during the design phase for the communication system also shown in Figure 3-17:

1. *Frequency Response*:   The frequency response of a filter defines which frequencies will be passed and which ones will be attenuated. The components that comprise the frequency response characteristics of a filter involve the passband, the cutoff band, the transition band, and the stop band.

   a) *Passband*:   This is one of the most important filter criteria. The passband defines which frequencies will pass through the filter and which frequencies will be discriminated against and attenuated. The passband is normally defined as the filter area which experiences the lowest level of attenuation, is ideally 0 dB and has a characteristic low-level ripple.

   b) *Cutoff Frequency*:   This is the frequency where the passband is at 3dB of attenuation or rather the end of the desired passband.

   c) *Transition Band*:   This is the portion of the filter's response that is between the cutoff frequency and the stop frequency. This is the part of the filter response where the greatest attenuation change occurs.

   d) *Stop Band Edge*:   This is the highest frequency at which the passband ripple occurs. The stop band edge is also the transition point where a small increase in frequency gives a large increase in attenuation. Receiver front ends usually specify the stop band

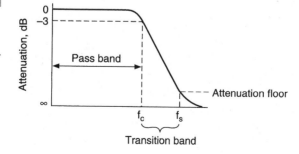

Figure 3-17
Filter frequency
response

attenuation at the upper and low frequencies it operates over. For a transmitter, the stop band frequency is normally specified with respect to the receive band for the receiver.

2. *Insertion Loss*:   The insertion loss or attenuation in the passband is important because it defines how much loss the filter will impose upon the signal as it transverses through the filter itself. Ideally, the insertion loss through a filter should be 0 dB. However, a passive filter will have some insertion loss and this insertion loss also equates to a noise figure for the filter itself.

In addition to insertion loss, there is also a factor called gain loss, which makes up part of the insertion loss that the filter imposes upon the signal. The gain error is the difference between the specified and actual passband gain, or insertion loss. For a typical filter, the gain error can be as high as a few percent. This value can be referenced to many things, including the frequencies below the cutoff for a low pass filter or the entire passband; it can even refer to the band reject portion of the filter. However, it normally refers to the entire passband. See Figure 3-18.

$$\text{Gain error} \ = \ 5\% \ = \ (1 - \text{dB insertion loss}) \times 1.05 \ = \ -1.05 \text{ dB}$$

$$\text{Insertion Loss (dB)} \ = \ \text{amplitude before filter} \ - \ \text{amplitude after filter}$$

3. *Passband Ripple*:   The passband ripple is the variation in gain, or insertion loss, over the passband; also referred to as the in-band variation of the signal. The plot of gain verse frequency response for the filter in Figure 3-19 shows a ripple across the passband instead of a flat response. Typically, a filter will have 1% to 2% ripple over most the filter's passband.

4. *Attenuation Floor*:   This is the highest attenuation level for the filter at the stop frequency. The attenuation floor reference, fs, is where the roll off response of the filter's transition band crosses the attenuation floor. The attenuation floor is referenced in Figure 3-17.

5. *Shape Factor*:   The shape factor for a filter is a measure of the filter's attenuation steepness for the transition band. The steepness factor will increase in value as the number of poles and/or zeros increases or as the order of the filter increases.

$$SF = fs/fc$$

Figure 3-18
(a) Insertion loss, (b) high pass filter

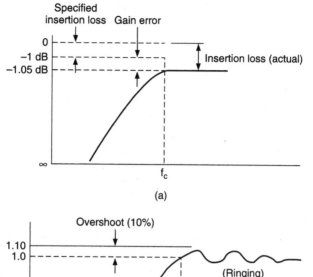

(a)

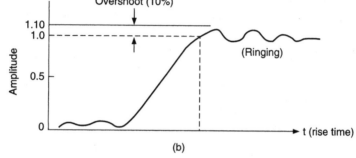

(b)

Figure 3-19
Passband ripple

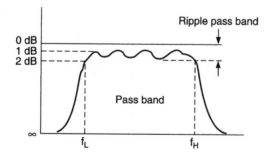

Ideally the steepness factor should be unity, $SF=1$.

6. *Phase Error*: Phase error is also referred to as phase linearity. Phase error is the linearity of the phase shift versus frequency. If there is no phase error, the phase linearity line should be a straight line. If there is no phase error, there is no group delay because the derivative a constant is 0 and the group delay is the derivative of the phase error.

7. *Group Delay*:   Group delay is defined as the time delay through a filter for a finite length of time for a burst (pulse). Ideally, the group delay for a filter should be constant across the entire passband of the filter. A group delay that is not constant across the passband can cause overshoot or ringing in the passband itself.

   The group delay for a filter is a derivative of the filter's phase shift and is mathematically represented as phase verse frequency response. The actual magnitude of the group delay is not important; rather, its flatness across the passband is the key to the filter's performance and is a measurement of distortion.

   If the group delay is not constant over the bandwidth of the passband for the filter where the desired modulated signal resides in frequency, some form of waveform distortion will take place. The narrower the desired signal in bandwidth, the less likely it will undergo any noticeable group delay.

   The tolerance in group delay over the entire filter's passband, sometimes called group delay ripple, is expressed in units of time such as milliseconds or microseconds. The absolute value is normally not mentioned, only the deviation from some fixed value.

   Finally, any impedance mismatch in the communication system will also invoke some group delay due to SWR problems.

8. Selectivity (Q):   The selectivity of the filter is another key attribute for the filter. The higher the selectivity of the filter, the better it rejects unwanted signals from being passed through the filter unattenuated. Ideally, the filter should be extremely selective and allow only the desired signals through without distorting the desired signal.

   The Q of a filter is defined as the ratio between the center frequency and the bandwidth of the filter.

9. *Temperature Stability*:   The temperature stability of a filter is a very important criterion. Specifically, the filter should be defined in terms of its tolerance in parts per million per degree of temperature change or ppm/C°.

   Ideally, a filter should retain its filter characteristics over the temperature range that it will be subjected to. But because the filter is constructed of various components that will change their physical dimensions, with temperature the chance for a change in frequency response for the filter is great. A common technique that is employed in the filter construction process is to utilize materials that will off set each other's change in physical characteristics as the temperature changes.

Receiver

The receiver and the receive system utilized by an LMDS system is a crucial element of the network. Specifically, the receiver's job is to extract the desired signal from the plethora of other signals and noise that exists in its environment. The basic receiver block diagram is shown in Figure 3-20 for reference.

The receiver involves the portion of the radio system including everything in the receive path, starting with the antenna system itself. When talking of a receiver in the more classical sense, it would involve only the portion of the network that is directly involved with the down converting and demodulating the signal to extract the initial information content. The rational, just like the rest of the receive system's components, is to treat it as a whole system because all the components in the receive path directly influence the ultimate performance of the receiver.

Types of Receivers

There are many types of receivers that are utilized for broadband wireless communication. The type of receiver utilized for the communication system should be selected so the information content to be received is done in the

Figure 3-20
Basic radio receiver

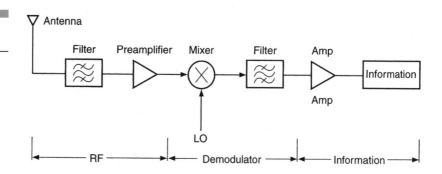

most efficient way. The most efficient method incorporates not only financial but also spectral efficiencies, where the desired information content is delivered to the receiver.

Normally, a communication system receive must deal with a signal spectrum that contains more than just the desired signal. The multitude of signals that the receiver must simultaneously deal with puts a price on device linearity for each of the stages in the receiver's path.

The receiver in its operation must select the desired carrier or signal, from a multitude of other signals, amplify the weak desired signal, and then demodulate it.

The top electrical performance and cost drivers for a receiver typically are as follows:

1. Frequency Range

2. Dynamic Range

3. Phase Noise

4. Tuning Resolution

5. Tuning Speed

6. Sensitivity

7. Distortion (gain and phase)

8. Noise

9. Others

The receiver design must incorporate the desired performance criteria and at the same time minimize the number of stages between the RF and IF portions of the receiver system. The number of stages between the RF and IF portions of the receiver system is dependant upon the modulation scheme selected.

The information content that is desired to be sent and received has a direct role in selecting the type of modulation format that will be utilized. The modulation format that will be utilized ultimately determines the type of receiver that will be utilized in the wireless network. There are many aspects to determining the type of modulation format desired in a network. However, one key criterion is the information bandwidth needed as compared to available spectrum. Specifically, if you have an infinite amount of spectrum to utilize, the type of modulation format chosen is done so based purely on the cost constraints imposed on the system operator for operating a profitable business.

However, you never have infinite spectrum, or bandwidth, to utilize for the wireless system. Therefore a tradeoff is needed to determine what the

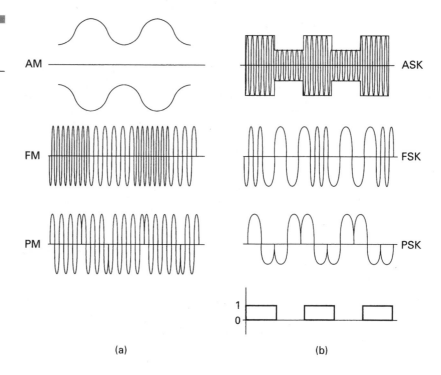

Figure 3-21
Modulation formats
(a) analog (b) digital

modulation format will be used to maximize the spectral and cost constraints of the system design.

The three general types of modulation types are shown in Figure 3-21 for reference for both analog and digital. Through manipulation of the previous equation, all the modulation schemes are achievable.

$$E(t) = A \sin (2\pi f_c t + \phi)$$

where A = amplitude
        f_c = carrier frequency
        ϕ = phase
        t = time
        E = instantaneous electric field strength

There are three general types of receivers: *Amplitude Modulation* (AM), *Frequency Modulation* (FM), and *Phase Modulation* (PM). Many receivers utilize a combination of the three types of basic receiver elements based on the technology platform utilized for the wireless system

Receiver System Blocks

The basic building blocks of a radio receiver system are shown in Figure 3-22.

The basic building blocks for a receive system are virtually the same for all types of modulation formats chosen. The chief differentiator for all types of receive systems is in the demodulation portion of the receiver itself. What this section of the chapter will do is cover the various aspects of the building blocks for a receive system.

Antenna System

The antenna system is the first stage in the receive path for the receiver. The antenna's purpose is to de-couple the electromagnetic energy from the atmosphere and transfer it to the feedline for the communication site. There are many attributes for a receive antenna to have (as discussed previously). However, for this part of the discussion it is assumed that the antenna system is properly matched and operates in the frequency band of interest with uniform gain of the whole bandwidth.

Feedline

The feedline is the physical device that connects the antenna to the rest of the receive system to the propagation medium that is used to transport the information. The feedline is comprised of cable and associated jumpers that normally connect the antenna to the receive filters. The feedline is an important element in the receive system and has a direct role in determining how well a receive system will operate.

The feedline referenced here, Figure 3-23, consists of a jumper cable, either LDF or Superflex, which connects the antenna to the coaxial cable, commonly known as the feeder or feedline. The jumper cable is used to connect the feedline to the antenna, purely because of physical constraints where the ability to bend a 7/8-inch cable is much more difficult than a 1/2-

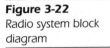

Figure 3-22
Radio system block diagram

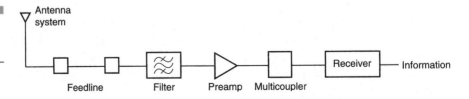

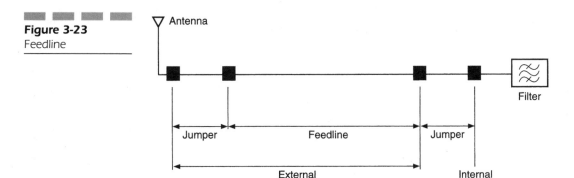

Figure 3-23
Feedline

inch cable. The jumper is meant to aid in the installation and maintenance portion for the antenna system.

Referencing Figure 3-23, the feedline usually enters the communication facility at the bulkhead for a shelter or external cabinet-type installation. The feedline is then terminated via a connector that again allows for another jumper cable to be installed. The jumper cable that is used between the feedline and the receive filters with the objective again is for ease of installation and maintenance of the facility.

The chief disadvantage with utilizing additional jumpers is the additional losses and potential connector problems that can arise for LMDS systems when the feedline connects the ODU to the IDU.

Filter

The filters utilized for a communication system are a key component of a communication system. There are normally several filters employed in a receive path for a communication system. Specifically, the filter should pass only the frequencies of interest, reject all the other frequencies, and do so with no loss in amplitude for the desired signals. Obviously, this is not practical to implement when space constraints and cost enter into the decision matrix. However, the filter that is used in the communication receive path has a large role in determining just how well the receiver will ultimately perform.

Preamplifier

The preamplifier is usually the first active component in a communication system's receive path. The basic function of any RF preamplifier is to increase the signal to noise ratio of the received signal. The preamplifier receives the desired signal at the lowest level in any of the receive stages for the communication site. Because the RF preamp receives the desired signal at the lowest level of any receive stage in the cell site's receive path, noise or other disturbances introduced in this stage have a proportionally greater effect.

The performance of the cell site's receiver with respect to weak signals depends on the performance of the preamplifier, or rather the signal to noise ratio of its output. The key issue is that amplifiers do not discriminate between what is the signal and what is noise or interference within the amplifier's passband. In fact, the preamplifier will amplify the desired signal plus any noise equally.

The preamplifier, as shown in Figure 3-24, sometimes has a degree of filtering incorporated into the preamplifier itself or is just a straight amplifier.

The preamplifier must have a sufficient gain to assist the receiver in its sensitivity. However, too much gain has the adverse effect of creating more intermodulation products when in the presence of strong signals. Additionally, any amplifier has a power budget, regardless of its gain and robustness. The preamplifier for the cell site will amplify the out of band emissions that made it past the cell site filters. Depending on the amount of the out of band emissions allowed to pass into the receive path for the cell site, a decrease in overall receiver gain can be experienced.

The preamplifier shown is an amplifier with redundancy built into the path. The objective with this design is to ensure that if one leg of the receive

Figure 3-24
Preamplifier

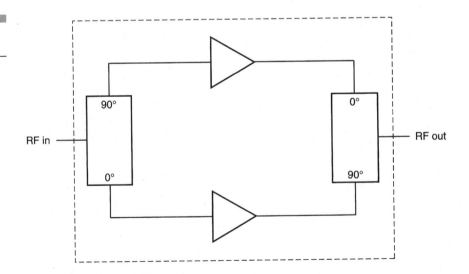

path is damaged for a multitude of reasons, the other path will ensure that the receiver signal is still allowed to pass to the rest of the receive system.

Multicoupler

The multicoupler is a device that ensures that received signals are routed to the appropriate receivers. Usually, the multicoupler has several stages of splitting, as shown in Figure 3-25. The multicoupler itself normally has the preamplifier included as part of the configuration, and is therefore included here. If the preamplifier is not included, the multicoupler contains only a combination of RF splitters.

The configuration shown enables one receive antenna to be connected to many radios. The key advantage with utilizing a multicoupler is the reduction in the amount of antennas and feedlines required for a cell site. By utilizing a multicoupler, many radios within a communication site can share the same receive antenna.

The multicoupler shown in Figure 3-25 implies that the receive path from the multicoupler itself to each individual radio is the same. However, if the path is different for say Radio 1 and Radio 16 because of internal RF plumbing issues, it might be necessary to include additional gain or padding for the receive branch or group of radios that has the imbalance.

Figure 3-25
Multicoupler

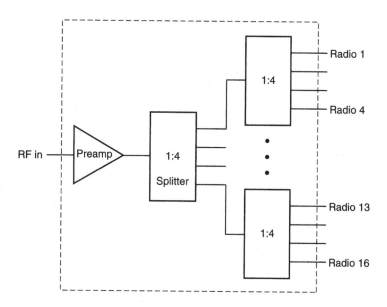

Depending on the LMDS system that is chosen, the use of a multicoupler may not be relevant if there is a one-to-one relationship between the antenna and the receiver.

Radio Receiver

The radio receiver of an LMDS device, whether located at the base station or the host terminal itself, is referred to here as the physical device that converts the RF energy into a usable form. The radio receiver can have from one to multiple receive paths connected to it. Usually, there are two paths connected to the radio receiver in a cell site and only one path for a mobile or portable unit.

The basic radio receiver is shown in Figure 3-26.

The radio receiver involves receiving the RF energy that is passed though a filter for additional selectivity and then amplified for additional gain. The RF energy is then put through a mixer, which enables the signal to be down converted to an intermediate frequency. The intermediate frequency is then filtered and amplified again.

The IF signal is now passed through another mixer and filter, which now places the IF signal into the audio or information stage. The audio or information stage is where the initial information content is conveyed to the desired receiver, either a person or electronic terminal device.

Obviously, the previous example is a simplified version of the event that takes place in the receive path. The specific sequence of events that takes place to the signal that goes through the receive path chain is dependent upon the type of modulation and information content that is trying to be interrupted by the end user or device.

Figure 3-26
Radio receiver system

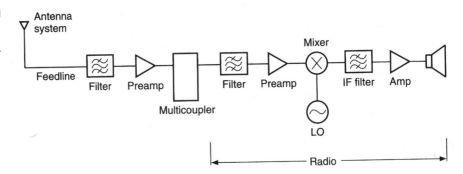

Receiver Filter   The filter in a receiver itself is meant to accomplish several functions. The first objective is to improve the radio's selectivity by having the filter eliminate all the unwanted energy that has been allowed to be passed and generated in the receive path of the communication system. The second objective, and the most important, is to protect the amplifier in the radio from desensitization and overload due to out of band emissions that could occur.

The filter that is used in the radio is a bandpass filter that usually can operate over the entire spectrum band of interest, and then some. The point that the radio receive filter is effectively wide places even more emphasis on the need to have a more selective filter at the beginning of the receive path system.

Preamplifier   The preamplifier for the radio itself is the first amplification stage in the actual radio itself. The amplifier in the receive path of the radio receiver's purpose is to help set the noise figure and sensitivity of the radio itself. In addition, the amplifier is meant to overcome any conversion losses experienced in the receive path as a result of the filter or mixer.

Mixer (Down conversion)   The receiver mixer has a critical role in converting the incoming RF spectrum containing the information content into a IF output, ideally without adding noise or intermodulation products along the way. In most modern radios, there are two mixers that are employed as part of the down conversion process. However, for simplification, only one mixer is shown.

Mixers are sometimes described in textbooks as multipliers. A mixer normally has three ports and is a vital component in either up- or down converting the information into another frequency. For a receiver, the mixer can either up- or down convert the frequency, but the standard method that is utilized is to down convert the frequency so it can be processed at an intermediate frequency band instead of the RF band.

The down conversion process for the receiver is usually accomplished through means of a mixer. The mixer is represented in system block diagrams as a circle with a cross through it. The mixer's function is to translate the incoming or source signal with that of another signal, usually the local oscillator frequency. A brief diagram of a mixer is shown in Figure 3-27 for reference.

In most receivers, there are two down conversion processes that take place: the first involves reducing the initial RF signal to a level that can be processed better in the receiver. This is accomplished by converting the RF

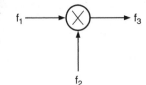

Figure 3-27
Mixer

signal to an *intermediate frequency* (IF) and then mixing it again to reduce it even further to a second IF level. The rationale behind this method is to improve the overall performance of the receiver.

For a mixer, it is very important to ensure that the device is matched to not only the f1 portion, usually the RF or source signal, and also to f2, the local oscillator port. Proper matching ensures that phase and gain matching over the desired bandwidth takes place without adding distortion and/or phase noise to the signal.

IF Stage   The IF stage of the receiver normally undergoes an additional down conversion, so the second intermediate frequency is at a lower frequency for ease of post processing. Most of the amplification in the receiver takes place at the IF level. The IF stage is an important part of the receive chain because this is where the post processing of the information takes place. The IF stage can take on several variants, depending on the technology selected for use. However the basic premise is the same in that the signal is now at a lower frequency range where it can be post processed easier.

There are several important criteria that define the process. They involve IF selectivity and image rejection (to mention two important figures of merit).

IF selectivity is probably one of the most important specifications for a receiver. To prevent interference between channels at the receiver, IF selectivity is used to obtain the necessary interference protection. The selectivity of an IF section is a measure of the total response of all the IF stages, if there are several involved with the process. The selectivity of the receiver must be sufficient to allow a desired modulated signal to be amplified uniformly across the desired band, but yet reject all the unwanted energy.

The selectivity of a receiver is usually defined in terms of its Q.

The image rejection is an unwanted signal specification. This is usually a filter specification and is the level in dB between the desired signal and the image signals power. It is measured by applying a signal that is at an image frequency and then increasing its signal strength until it is detected.

Then, the desired signal is applied until it is detected. The difference is the image rejection level.

$$\text{Image Rejection} = \text{P received} - \text{P image detected}$$

$$\text{where } \text{P received} = \text{P desired} = \text{P image before filtering}$$

Ideally, the image rejection should be infinite, or rather the same value as the power received, meaning that there is no image energy detected. Remembering the image example presented before, the source signal will also produce an image that falls within the passband of the receiver. IF filters should be attenuated by either the cell site filters and or the receiver filter. The attenuation of the signal and the selection of the local oscillator should be selected to ensure that no adverse problems occur in the receiver itself.

Information Stage   The information stage of the receive process is where the initial information content is extracted and utilized by the terminal or users at the end of the communication link. The audio stage is where the signal is eventually demodulated and transformed into information. The demodulation of the signal is dependent upon the modulation and information content desired for the LMDS system.

Performance Criteria

The performance criteria for a radio receiver are the key elements for determining if the radio selected to extract the information content into a usable form will do so with success or failure, depending on the environmental conditions it is placed within.

The performance criteria covered in this section do not pertain to the physical environmental issues, power consumption, heat exchange requirements, and *mean time between failure* (MTBF). Instead, the objective of this section is to provide a reference to the radio performance criteria that a design engineer should utilize in selecting a radio for the network.

Sensitivity

The capability for a receiver to detect a weak signal is determined by its sensitivity. Receiver sensitivity is a very important FOM for a receiver. The receiver sensitivity must be such that it can detect the *minimal discernible signal* (MDS) from the background noise. The MDS is a measure of sensitivity that incorporates the bandwidth of the system. The MDS will differ from one receiver to another, based on the bandwidth of the signals received.

Specifically, there is a relationship between thermal noise, the receiver's noise figure, and the bandwidth of the signal that the receiver is trying to detect. The relationship for receiver sensitivity is defined as follows:

$$\text{Sensitivity} = 10\log(kTB) + 10\log(\text{bandwidth(Hz)}) + NF \, (\text{dB})$$

$$= -174 \, \text{dBm/Hz} + 10 \, \text{Log(bandwidth(Hz)} + NF \, (\text{dB})$$

$$k = \text{Boltzman constant} = 1.38 \times 10^{-23 \, \text{J/K}}$$

$$T = \text{temperature in } K$$

$$B = \text{Bandwidth (Hz)}$$

Selectivity

Receiver selectivity is a measure of the protection that is afforded the radio from off-channel interference. The degree of selectivity is largely driven by the filtering system within the receiver. The IF portion of the receiver affords the most benefit for selectivity. The greater the selectivity, the better the receiver is able to reject unwanted signals from entering into it. However, if the receiver is too selective it may not pass all the desired energy.

Dynamic Range

The dynamic range is a very important FOM and it defines the range of signals that the receiver can handle within the specified performance of the receiver. There are several ways to specify the dynamic range of the system. One way is to define it as the range from the MDS to the 1 dB compression point of the receiver. This is often called the blocking dynamic range. Another method of defining dynamic range is to define the range from the MDS to where the 3rd order IMD equates the MDS signal. This is referred to as *Spurious Free Dynamic Range* (SFDR), or it can be specified as the difference between MDS and a specified IMD level.

Figure 3-28 shows a chart that can be used to determine the dynamic range of a radio system. Please note that the signal slope in the chart has a

Figure 3-28
Dynamic range

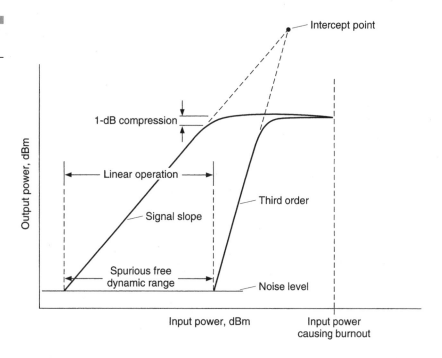

slope of 1:1, whereas the 3rd order intercept has a slope of 1:3. The actual intercept point referenced on the chart is a calculated value only.

SFDR is a very important specification when the site is near other radio transmitters because it is a direct indication of how the signal interferes with adjacent channels. SFDR provides a measurement of the radio performance as the desired signal approaches the noise floor of the receiver. It provides an overall receiver SNR or BER; for example, if a radio can accurately digitize signals from -13 to -104 dBm in the presence of multiple signals, the DR $= -91$ and it implies a SFDR of 95–100 dB.

The SFDR can be improved when the signal level is reduced from the full scale and this can improve the actual dynamic range, even with the reduction in signal amplitude.

Dynamic range is usually defined as the range over which an accurate output will be produced. The lower level is called sensitivity and the upper level is called the degradation level. The sensitivity is determined by noise figure, IF bandwidth and the method of processing. However, the degradation level is determined by whichever component in the receiver reaches its own degradation level first. Therefore, it is important to understand all the components in the receiver path because the first component to degrade will define the dynamic range of the system.

A large dynamic range for a receiver is a design priority. When comparing receiver to receiver, assuming that they are operating in the same band and have the same bandwidth for receiving, the following FOM can be defined to compare one receiver to another:

$$\text{Dynamic Range FOM} \;=\; \text{Input IP3} \;-\; \text{Noise Figure}$$

If the output IP3 is referenced for the receiver, then the Input IP3 can be calculated once the gain of the device is known.

$$\text{Input IP3} \;=\; \text{Output IP3} \;-\; \text{Gain of Device}$$

The Dynamic Range FOM should always be referenced to the system input; by using this method, a high dynamic range usually results in a positive FOM for the receiver.

An example of how dynamic range is calculated is shown in Figure 3-29.

Figure 3-29
Dynamic range

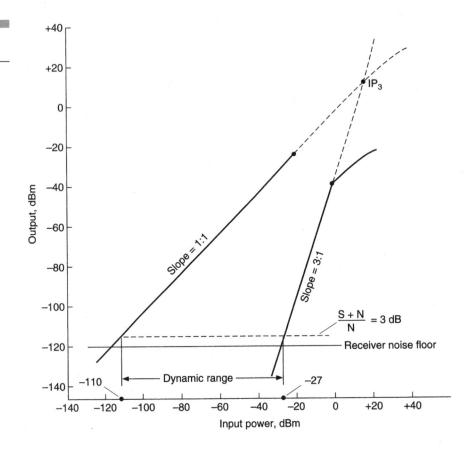

Distortion

Distortion simply means that unwanted signals appear at the output of any device in the RF path. A common place for distortion to occur is at the output of an amplifier (for this discussion, it will be the preamp for the communication site).

Distortion generally takes on one or more forms; the three most common forms are harmonic distortion, intermodulation distortion, and cross-modulation distortion. Whenever distortion occurs in the original signal, some level of degradation occurs. Depending on the severity of the distortion, the communication content may or may not be disturbed.

Harmonic distortion occurs when the unwanted signals from multiple carriers are some integer multiple of the initial signal. For example, an amplifier boosts a signal at 25.5GHz, and at the same time generates a harmonic distortion starting at 51Ghz for the second harmonic and 76.5GHz for the third harmonic. This has the negative effect of robbing the energy dedicated to the initial channel and creating unwanted signals in other parts of the spectrum.

Crossover distortion occurs when the amplitude modulated signal from one transmitter is transferred to another carrier at the output of the device.

Intermodulation distortion is the most common form of distortion and is the product of several signals mixing together. The amount and levels of *intermodulation distortion* (IMD) are a direct result of the amount of signals that are available to be mixed at any location. Intermodulation products for a second- and third-order mix for two signals are shown in Table 3-1.

If f1 = 25GHz, f2 = 26GHz.

The intermodulation table for 2nd and 3rd order products is shown in Table 3-2.

Regardless of the type of distortion, the location for its occurrence can be at the audio or baseband prior to modulation, during demodulation, or somewhere at the RF level after modulation.

	2nd Order	3rd Order
Table 3-1		
IMD	A+B	2A+B
	A−B	2A−B
	B−A	2B+A
	2B−A	

Table 3-2

IMD products

Order	Mix	Product (GHz)
A+B	25+26	51
A−B	25−26	NA
B−A	26−25	1
2A+B	2(25)+26	76
2A−B	2(25)−26	24
2B+A	2(26)+25	76
2B−A	2(26)−25	27

Noise

Noise for a communication system directly affects its overall performance. All receivers need to have a certain C/I or Eb/No value to perform properly. If the overall noise that the receiver experiences or has to deal with increases, the desired signal needs to be increased in signal without increasing the noise content, to ensure that the proper ratio is maintained.

For the RF design engineer, there are several components associated with the receive system that comprise noise: thermal, shot, and system noise. The latter can be reduced through use of proper frequency planning, power control, and appropriate use of selective filters along with isolation techniques. Both the thermal and short noise comprise what is referred to as the noise figure for a receiver.

Noise Figure Noise figure is one of the fundamental measures of a receiver's performance and should be measured at a predetermined location for the receiver itself. Noise figure for a receiver degrades (that is, it increases) with each successive stage in the receive path. A common point to measure noise figure is the audio output for a receiver, but with digital radios there is no audio output and the measurement point is then the IF output.

The noise figure of the receive system is directly related to the overall receiver sensitivity. The noise figure for the system is calculated as follows:

$$\text{Noise Figure} \; = \; FN \; = \; (\text{Sin}/\text{Nin})/(\text{Sout}/\text{Nout})$$

$$\text{Noise Figure} \; = \; FN \, (\text{dB}) \; = \; 10 \log (FN)$$

The noise figure for the system is normally set by the first amplifier in the receive path. The noise floor can be improved or attenuated by passive devices, but the SNR will not be improved unless the bandwidth is narrowed.

Thermal and Shot Noise   There are two components to noise: thermal and shot. Thermal noise is a direct result of kinetic energy and is directly related to the temperature. Shot noise is caused by the quantized and random nature of current flowing in a device. Both thermal and short noise comprise the noise that exists in a communication system.

Noise temperature has a direct relationship to operating frequency and the bandwidth of the receiver or signal desired to be detected. The relationship is shown in the following equation:

$$\text{Noise Power per Hertz} = 10\log(kTB)$$
$$= -173.97 \text{ dB/Hz (25C or 290K)}$$

What is of prime interest to the RF engineer is what effect each component in the receive system has toward improving the overall sensitivity of the system. The first amplifier in the system sets what the noise figure for the system will be. The equation to determine the noise figure is shown next:

$$F = F1 + ((F2 - 1)/G1) + ((F1 - 1)/(G1 \times G2))$$

Using the previous example, however, a quick relationship can be determined between noise figure and receiver sensitivity.

$$\text{Sensitivity (optimal) dB} = -174 \text{ dB} + 10\log(B) + NF$$
$$B = \text{bandwidth (Hz)}$$
$$NF = \text{noise figure}$$

The previous calculations, however, do not factor in the effects of the antenna noise, feedline loss, and required S/N or Eb/No as part of the receiver sensitivity.

dB Compression

The 1 dB compression point is a common reference term used to define the performance of a particular receiver, or the amplifier in the receiver itself. When the 1 dB compression point is referenced for the entire receiver, the

1 dB compression point can also reference an individual amplifier such as the preamp.

The 1 dB compression point is where the power gain for the receiver is down 1 dB from the ideal gain. That is, if the input signal goes up by 2 dB and the output goes up by only 1 dB, this is the point where the 1 dB compression point occurs.

Often, the 1 dB compression point is referred to as blocking for the receiver. The blocking occurs in that the weaker signals are not amplified properly, leading to them being potentially blocked from being detected. The 1 dB compression point is part of the component for determining the receiver's overall dynamic range.

The 1 dB compression point is shown in Figure 3-30.

The 1 dB compression can occur either as a direct result of trying to receive the desired signal, which is too hot causing the overload condition or through undesired signals overloading the receiver.

Referring to Figure 3-30, above the 1 dB compression point can be directly affected by the receivers overall gain setting.

Third Order Intercept

The third order intercept point (IP3) is a FOM for a receiver. Specifically, the IP3 value determines directly influences and determines the receiver's

Figure 3-30
dB compression point

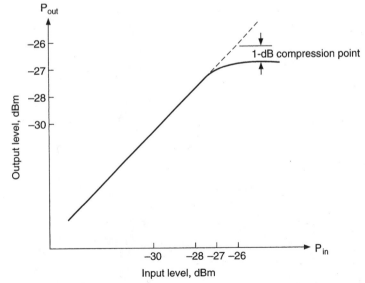

dynamic range. The 3rd order and 2nd order intercept value determine the receiver's linearity.

It should be noted that the IP3 value is a theoretical value that is achieved due to extrapolation of the 3rd order curve. In addition, the IP3 value is frequency dependent and when based on the selection of the test frequencies, a different IP3 value may be achieved. However, most IP3 tests are set up to produce a product that falls within the first IF of the receiver.

Basically, the IP3 value increases by 3 dB for every 1 dB increase for the desired signal. The slope of the IP3 line is therefore 3:1 and is shown in Figure 3-28. The IP3 value is important for determining the receiver's performance because the presence of two larger signals can generate spurs caused by non-linearities, or distortions, that can override the weaker desired signals.

Phase Noise

Phase noise is a measure of the receiver's phase and frequency perturbations that are added to the initial input signal. The effect of phase noise is to distort the initial signal, which either degrades the initial signal or degrades the receiver's sensitivity.

Phase noise is a specification of spectral purity and is usually defined at $-dBc/Hz$, which means that at a certain offset from the center frequency, the phase noise must be x dB below the center frequency. The phase noise is commonly specified as SSB noise spectral density.

Phase noise is an important FOM because it can lead to frequency translations occurring for out of band signals to fall within the receiver's passband.

Desense

Any form of RF energy, whether it is manmade or natural in origin, has the potential of adversely affecting the receiver. Any RF energy that adversely affects the receiver's ability is defined as RFI. Desense is technically a reduction in the receiver's overall sensitivity, which is caused by RFI, either manmade or natural. Desense occurs when a very strong signal begins to overload the front end of the receiver and makes the detection of weaker signals more difficult.

All types of noise interference adversely affect the cell site's receiver performance, whether the desired signal is purely FM analog or TDMA. Noise interference can either be attributed with out of band emissions from nearby transmitters whose side bands contribute to the effective noise floor of the cell site or it can come from inband sources, cochannel, or IMD products.

The magnitude of the noise interference is determined by the amount of energy within the receiver passband. Depending on the amount of noise interference, the effective receiver sensitivity can be degraded during increases in the noise interference level because of reduced S/N or BER ratios.

When colocated with many transmitters, the transmitter noise spectrum is noticeably close to the carrier at 60 dBc and gradually falls off in power as a function of effective bandwidth. However, at some power level, the noise level will degrade receiver sensitivity. The noise interference power level caused by other transmitters can be reduced by increasing the physical separation between the offending transmitter(s) and the offended receiver(s). Another alternative is to place a more narrow transmit filter on the transmitter or utilize a more selective receive filter to remove or seriously reduce the transmitter noise spectrum.

Whenever transmitter noise is present, there exists the potential for receiver desensitization. The cell site receiver can be desensitized by undesired signals reducing the sensitivity of a receiver by reduction in the gain or an increase in noise level of certain amplifiers or mixers in the receive path of the cell site. It is possible, based on the receiver filter used, to pass out of band energy into the receive system and cause a reduction in receiver gain. The loss of gain in this manner is a form of desensitization.

Therefore, if a cell site is desensed due to out of band emissions, the installation of a more selective filter will make the site more sensitive. However, if the site is not being desensed due to out of band emissions, the filter's introduction could show up as a reduction in effective noise floor for the cell site.

The basic function of any RF preamplifier is to increase the signal-to-noise ratio of the received signal. Because the RF preamp receives the desired signal at the lowest level of any receive stage in the cell site's receive path, any noise or other disturbances introduced in this stage has a proportionally greater effect.

The performance of the cell site's receiver, with respect to weak signals, depends on the performance of the preamp or rather the signal-to-noise ratio of its output. The key issue is that amplifiers do not discriminate between what is the signal and what is noise within their passband. In fact, the preamp will amplify the desired signal plus any noise equally.

Any amplifier has a power budget, regardless of its gain and robustness. The preamp for the cell site will amplify the out of band emissions that made it past the cell site filters. Depending on the amount of the out of band emissions allowed to pass into the receive path for the cell site, a decrease in overall receiver gain can be experienced.

The best method to overcome desense problems is to ensure that there is sufficient isolation between the potential offending transmitter and the receiver. Desense will continue to be more of a problem, especially in urban environments, as the use of wireless communications continues to expand.

References

American Radio Relay League. *The ARRL Antenna Handbook*. 14th ed. Newington, CN: The American Radio Relay League, 1984.

American Radio Relay League. *The ARRL Handbook*, 63rd ed. Newington, CN: The American Radio Relay League, 1986.

Breed, Gary. "Receiver Basics-Part 1: Performance Parameters." *RF Design*, February 1994, 48–50.

Breed, Gary. "Receiver Basics-Part 2: Fundamental Receiver Architectures." *RF Design*, March 1994, 84–89.

Carlson, A. Bruce. *Communication Systems*. New York, NY: McGraw Hill, 1996.

Carr, Joseph J. *Practical Antenna Handbook*. New York, NY: McGraw-Hill, 1989.

Christiansen, Donald. *Electronics Engineers Handbook*, 4th ed. New York, NY: McGraw-Hill, 1996.

Fink, Donald G. and H. Wayne Beaty. *Standard Handbook for Electrical Engineers*, 14th ed. New York, NY: McGraw-Hill, 1999.

IEEE Standard 184-1969, "IEEE Test Procedures for Frequency Modulated Mobile Communication Receivers." New York, NY: Institute of Electrical and Electronics Engineers.

Jakes, William C. and Donald Cox. *Microwave Mobile Communications*. New York, NY: Institute of Electrical and Electronics Engineers, 1994.

Johnson, Richard C. *Antenna Engineering Handbook*, 3rd ed. New York, NY: McGraw-Hill, 1994.

Kaufman, Milton and Arthur H. Seidman. *Handbook of Electronics Calculations*, 2nd ed. New York, NY: McGraw-Hill, 1988.

Lathi, B. P. *Modern Digital and Analog Communication Systems*, 3rd ed. New York, NY: Oxford University Press, 1994.

Lee, William C. *Mobile Cellular Telecommunications*, 2nd ed. New York, NY: McGraw-Hill, 1996.

Pawlan, Jeffrey. "A Tutorial on Intermodulation Distortion: Part 2-Practical Steps for Accurate Computer Simulation." *RF Design*, March 1996, 74–86.

Rappaport, Theodore S. *Wireless Communications: Principles and Practice*. New York, NY: Prentice Hall, 1995.

Schwartz. *Communication Systems and Technologies*. New York, NY: Institute of Electrical and Electronics Engineers, 1996.

Smith, Clint and Curt Gervelis. *Cellular System Design and Optimization*. New York, NY: McGraw Hill, 1996.

Smith, Clint. *Practical Cellular and PCS Design*. New York, NY: McGraw-Hill, 1997.

Steele, Raymond. *Mobile Radio Communications*. New York, NY: Institute of Electrical and Electronics Engineers, 1992.

Stimson, George W. *Introduction to Airborne Radar*. El Segundo, CA: Hughes Aircraft Company, 1983.

Van Valkenburg, Mac E. *Reference Data for Engineers: Radio, Electronics, Computers and Communications*, 8th ed. Newnes, 1996.

Watson, Robert. "Guidelines for Receiver Analysis." *Microwaves & RF*, December 1986, 113–122.

White, Duff. *Electromagnetic Interference and Compatibility*. Gainesville, GA: Interference Control Technologies Inc., 1972.

Williams, Arthur B. and Fred Taylor. *Electronic Filter Design Handbook*, 3rd ed. New York, NY: McGraw-Hill, 1995.

Yarborough, Raymond B. *Electrical Engineering Reference Manual*, 5th ed. Belmont, CA: Professional Publications, Inc., 1990.

Business
Considerations

Introduction

The business considerations necessary for an LMDS or any wireless system are as diverse as the amount of technology platforms and bands of operation that exist for this industry. The business considerations weigh heavily in the deployment and implementation decisions for any company. For example, what are the services that will be or can be offered? This is a simple question, but it is often extremely difficult to answer. The services selected should match the market that is being sought, which is obvious. However, just how will the services decided upon be distributed, that is, will they be selective or global? Will the services offered be selectively targeted only to islands of interest or will the service be advertised for the entire market? There are technical and financial issues with both of these approaches.

For instance, if the decision is for an island approach, it involves more of a direct sales force and a priori knowledge of from where your business originates. Specifically, a single base station may be constructed that covers an industrial park or residential community. With the island approach, that area is the only one that can be targeted for services until additional stations are installed.

Alternatively, if the decision is to roll out the service for the entire market, the decision must be made about how to handle requests for service where base stations are not deployed yet. More to the point, as has been witnessed with several LMDS systems, they declare service for an entire market, but in reality only limited areas are served. This creates the issue of turning away potential customers that are seeking your services based on what was advertised, or implied.

There are positive and negative consequences with both the selective and global service offerings. Therefore, a middle ground between these two methods can be achieved. For instance, there can be a limited roll out covering a specific market area, more than one site, and direct marketing toward that area, with a global ad campaign indicating where service is being offered and expected future offering dates, indicating a system rollout strategy. The beauty of LMDS in this approach is that you can target the areas where there is a match between the marketing plans with real data collected by potential customer inquiries for when service would be available in their area.

The primary objective with an LMDS is to make money. The governing method for determining the money made for the LMDS system is driven by the following equation:

Gross Profit = Gross Revenue − Gross Expenses

Although this may seem primitive in nature, the fundamental issue is that this concept is not fully understood. And you can never underestimate that your competitor is also aware of this concept. The following two brief examples are used for simplification to drive this point home.

For instance, if the business being pursued is Leased Line Replacement, the following is used:

$$\text{Gross Revenue} = \# \text{ T1/E1} \times \text{Revenue/year/T1(E1)}$$

$$\text{Expenses} = \text{Operating Expenses} + \text{Depreciation of Capital Equipment}^*$$

* For the example, depreciation is left out in addition to taxes, cost of capital, etc.
Therefore:

$$\text{Revenue/year/T1(E1)} = (\$250/\text{month} \times 12) = \$3000$$

$$\text{Operating Expenses} = \$5 \text{ million}$$

Which means to break even on cash flow, not profitability, the following T1/E1s need to be sold:

$$\#\text{T1/E1} = (\text{Operating Expenses})/(\text{Revenue}) = 5,000,000/3000 = 1667$$

However, the timing of this is not correct because it assumes that all of the T1/E1 are in service for the entire year and not added to the system over the course of the year.

Using the same information used previously for operating expenses (which is not valid due to different operating requirements), if the decision is to offer residential IP Always-On service, then the following can be extrapolated.
Therefore:

$$\text{Subscriber Revenue/Year } (\$20 \times 12) = \$240$$

This means that to break even on cash flow, not profitability, the following residential subscribers are needed for the same operating expenses:

$$\# \text{ IP Subscribers} = (\text{Operating Expenses})/(\text{Revenue})$$
$$= 5,000,000/240 = 20,834 \text{ Subscribers}$$

As with the T1/E1 example previously presented, the timing of subscriber revenue is not correct, but the examples shown illustrate an important point: the services offered and the take rate for those services need to provide more revenue than the cost of delivering those services.

However, you might make the decision that in order to capture market share, the service provided may initially be more costly to deliver than the revenue it will generate. For instance, to capture a customer for an area that does not have LMDS service yet, the choice may be made to utilize the PTT for providing the connection as a method for expediting the sales take rate and customer base.

Finally, it needs to be stressed that with the exception of a greenfield system, the primary competitor is the PTT in most markets, at least initially. However, the concept of cutting the price of the PTT as the primary business strategy will have short-term advantages, but will ultimately fail because the cost of an abandoned plant by the PTT will always allow the PTT to exercise the last comment on the price war. Instead, the golden path of LMDS is services brought upon by the bandwidth that is now made available through this exciting technology platform.

Marketing

What the company is today and what it will become when it grows up is the profound question that the marketing department needs to provide. The vision for the company and how it envisions where the revenue will come from is the billion-dollar question that they have to answer.

The marketing department's decision for what services will be offered (SLAs), where they will be offered, and the time frames for the offerings need to be defined with a degree of specificity. Obviously, when putting together a plan, the exact details or information may not be readily available but the decision to offer IP or Leased Line Replacement will have a fundamental implication regarding the infrastructure chosen and the support of the SLAs. For instance, if the decision is to offer Leased Line Replacement, but the equipment can only support IP through a 100/10 Base T connection, the platform needs to be revised. In addition, if the decision is to provide Always-On IP with SLAs that define a CIR of 512K (for example) for each connection, the bandwidth available from the radio environment plus the infrastructure's ability to ensure this offering because of over-subscription needs to be verified.

More specifically, the marketing group needs to work with the technical community to review that the vision regarding the market needs to be married with the technical community prior to the infrastructure decision. In addition, after the technology platform has been chosen, the two groups need to work together for the sole purpose of maximizing the market potential by matching the infrastructure's capabilities with the target market. To make this work, the marketing department needs to have a fundamental technical understanding of the target market, the barriers for offering the services desired, the infrastructure used, and how the services will actually be delivered to the customer.

The decision to be a broadband provider for large businesses, SMEs, SOHOs, residential, some blend in-between, or everyone needs to be decided from the beginning. As stated previously, you need to determine what you want to be when you grow up and not just decide to go for everything and let the market decide. Additionally, the services and their take rate must provide more revenue than they cost to deliver.

The following is a brief outline of some of basic marketing steps that need to be taken. Often, specific market information is not available in traditional methods or is too cost-prohibitive. A key concept that is often overlooked is that all too often your best source for information is your competitor.

The basic steps are the four basic steps that govern marketing. They are referred to as the four Ps: product, place, promotion, and, of course, price.

Product. This involves defining services offered, installation, warranty, product lines, packaging, and branding.

Place. This refers to the marketing objectives (penetration), the channels used for achieving the penetration (direct and indirect), market exposure and competition (PTT, CLEC, xDSL), distribution methods (direct and indirect), VAR partnerships, distribution of add-on services and products, and sales locations such as direct and indirect stores.

Promotion. This basically involves the bundling or unbundling method, direct and indirect sales force (number, training, incentives), advertising (media type, copy trust, trade shows, press, ad types, agency, if used). Also included is the sales promotion that involves commissions, customer retention programs, discounts, and VAR products.

Price. This involves the objective, which can be to ensure profitability for every product offering or have a loss leader in the mix. Also included are flexibility in terms of pricing for volume purchases or bundling techniques, product life cycle pricing, local market pricing differentiation, and sales allowances for indirect sales or CPE replacement in conjunction with the host terminal.

All four Ps center around the important issue of enticing the customer to want and perceive value with the product, use the service, and continue to use the service and possibly utilize additional services either offered now or some time in the future.

From the inception, the marketing department needs to feed many groups within the company. The sales department obviously needs to know what the vision is and the product offerings that will be offered for initial sale. The technical community needs information related to traffic or capacity planning to ensure that the factory is in place for the sales force to sell, while minimizing the capital and operating expenses. The customer care, billing, and provisioning groups need to know the market and services so that bill can be issued. And, of course, the customer must be serviced in as high a quality as economically possible, based on the services offered.

Marketing information is not static, but is dynamic in that market conditions always change and the company must respond to those changes, the response could be not to act. Therefore the marketing department needs to interact with the other functioning entities within the LMDS company to ensure continued success and refinements take place.

From the technical communities' aspect, the marketing information will drive the infrastructure layout and design. The RF and network design aspects are not the same, but they utilize the same fundamental information to derive their answers and designs.

Therefore, to properly project the capacity-handling requirements of a system, marketing plans need to be factored into the design. The marketing department's plans are the leading element in any system design, both RF and network. The basic input parameters needed from the marketing department are listed as follows. The information is critical, regardless of whether the system is a new or current system.

- Identification of key coverage areas in the network for system turn-on, or enhancements

- The projected subscriber growth for the system over the time frame for the study

- Projected CBR, VBR, and UBR traffic; and types of service offerings per quarter

- Demographic breakdown of customer types, services, and their logical locations

- IP and circuit switch traffic blends with potential migration percentage of existing customers

- Edge terminal location multiplication factor (how many customers per edge terminal)

- Special promotional plans over the study time period, such as all you can eat Internet at 512kbps

- Top 10 potential sales areas for the system (they should match the demographic data provided)

There is a multitude of other items needed from the marketing department for determining the system design. However, the information on the basic topics listed previously will be enough to adequately start the system design.

Additionally, if the system is new, then the information needs to be provided upward to nine months prior to implementation of the system. If the system is in operation, the data needs to be reviewed on a quarterly basis to ensure that there are no major design variations.

Services

Services is a vast area that has multiple meanings. For instance, the service offering could be for wholesale service providing only (that is, the pipe provider). Another example could be a retail service provider who provides the service directly to the customer. But what exactly is service? Is service the delivery of the bandwidth only? Or do you also provide adjunct services to support the primary bandwidth provision (sell CPE, cable the customer, configure their routers, provide ASPs and so on)?

Often, the services offered will change with time due to the varying market conditions brought on by the competition, which the LMDS operator is. However, you must decide what you will present to the customer. More specifically, if you are just a pipe provider, how do you ensure that they have been connected properly and, in the event of a problem, will receive the required support in an expedited manner?

The appearance of the company plays an equal role in determining the satisfaction of the customer, in addition to the physical services they are paying for. Everyone has examples of how opinions of a company were enhanced based on how well they were treated. There are more examples of negative images of companies due to how the customer was treated when a problem arose. The problem could be in the delivery of the service, the time it took, the forms required, the appearance of the install team, billing inquiries, and recommendations for more economical solutions or service offerings.

The services chosen to be transported over the network need to be able to be supported by the network. Although this seems obvious, if the decision is to offer T3/E3 service to a customer but the bandwidth is only 12.5MHz for the radio, it may be possible but the edge terminal interface needs to support this effort. However, based on LMDS spectrum allocations, a customer requesting T3/E3 service should be supported with a PtP microwave link, using a different band. The point is that the service offered should be reviewed by the technical community to ensure that there are no problems with supporting that service.

Another simple issue is when an SLA that the customer wants requires a 99.999% network reliability, but the interface to the PTT provides only 99.5% reliability. It may represent a problem because the weakest link in the chain determines its strength.

The following are just some types of services that can be considered. It is important to stress that the fact a service exists in the market does not mean it needs to be offered by the company itself. Another important aspect is that some services may be inefficient on the bandwidth, spectrum, allocated for use either due to physical channel bandwidth or the technology platform chosen. A solution to this situation could be to offer the service, but to do so at a price that is not competitive with the market, thereby offering it but reducing the potential take rate via price.

Bandwidth

1. Leased Line (T1/E1)
2. Fractional T1/E1 (n*DS0)
3. Frame Relay (n*64)

   - PVC
   - SVC

4. ATM (n*512k) rt-VBR
5. PRI
6. ISDN (2B+D)
7. X.25 (n*64)

Services

1. Internet Connectivity
2. Always-On
3. Virtual ISP
4. Web Hosting, E-mail, DNS
5. E-Commerce
6. Leased Line Replacement
7. Frame Relay

8. X.25

9. ISDN

10. Call Treatment

11. Switch Partitioning

12. Centrex

13. Long Distance Call Delivery

14. International Call Delivery

15. VPN

16. LAN-LAN

17. WAN

18. Video Conferencing

19. VoIP Gateway

20. FaxIP Gateway

21. Application Service Provider (vast area of potential VAS)

22. Billing

23. Network Monitoring

24. IT Department Replacement

25. Computer Rental

26. Customer Premise Wiring

27. Computer Installation/Upgrade/Software Sales

Just which services will be offered and to whom is again directly dependent upon the type of customer sought. Figure 4-1 is a simple chart, indicating the relative location each type of customer is with respect to each other.

SLA (Service Level Agreement)

Committed Information Rate (CIR)

QoS

Fixed or Variable Billing

Service Availability (99.XX%)

The SLA list provides the concept that based on the services offered by the company, different committed, guaranteed, bandwidths will be offered to the customer based on the issue of whether the traffic is CBR-, VBR- or UBR-based. The QoS levels are also associated with the CIR, which follow hand-in-hand with the CIR offered and the network service reliability.

Figure 4-1
Customer pyramid

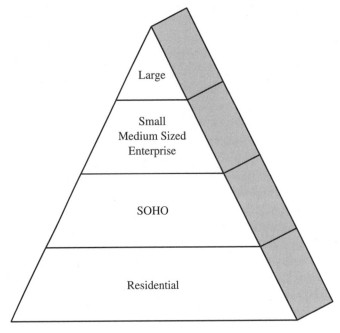

Customer Type Location
Relative to Market Size

From the previous list, which is not all-inclusive, the determination about whether you are to be a bandwidth provider or a service provider needs to be decided up front. If the decision is to be a full-service provider, the choice of which services to offer needs to be seriously considered. All too often, it is easy to list a potential service on a PowerPoint slide, facilitating a disconnect between the desired offering and the company's core competency. For instance, if the decision is to provide computer maintenance for the customers but the core competency of the company is to be a network pipe provider, the potential for less than optimal performance then exists. However, the lack of a particular core competency can always be resolved through the use of acquisitions and mergers.

Dual Band Strategy

A dual band strategy is more applicable to the 24GHz, 26GHz, 39GHz, and 40GHz operators. In particular, the use of a dual band strategy involves using a lower band in conjunction with the LMDS frequencies offered.

Without repeating parts of the RF design portion of the book, the basic concept is as follows. Offering service to the entire (or a large part of) the license area (market) will in all likelihood be very capital-intensive and not produce the desired revenue required to justify that expenditure required. Therefore, a lower band can be used to provide the initial service for the area in question. After a predetermined level of customer usage density is achieved for a given area, the implementation of a higher band LMDS base station can be deployed.

Now, which lower band frequency can you use? Well, one that can be used is the UNII band, which would have an exceptional reach over higher LMDS bands. The problem with using the UNII band is that this is a shared band and interference could present a problem regarding the different SLAs offered. In addition, the UNII band is more suitable for IP traffic. The use of 3.5GHz in Europe, coupled with the 26GHz or 40GHz, is yet another combination that is possible. However, the 3.5GHz is a licensed band and would require ownership of that band or partnering with another operator.

Either way, the dual band approach is one that can minimize the capital infrastructure required for initially securing market share.

Traffic

The calculation of how the traffic volume and types of traffic that will need to be transported on the network is essential for the proper dimensioning of the system. The traffic data will need to be defined into types of service classifications. Traffic information needed by the design teams involves both of the following:

Total Traffic (TT)

Busy Hour Traffic (BHT)

The reason for the two different levels involves different uses for the data. *Total Traffic* (TT) is meant to provide information regarding the volume of traffic and type that the LMDS operator will process and treat. The TT will primarily be used for determining the total volume of traffic that will be delivered to the PTT or another CLEC for call delivery for off-net traffic. Knowing the TT expected, by type, the ability to obtain lower interconnect and delivery costs can then possibly be experienced for off-net traffic.

The *Busy Hour Traffic* (BHT) is meant to dimension the heaviest load by total and by service classification for each of the network nodes and its related infrastructure. The BHT will also determine the CIR that will be sought for connecting to other operators for off-net traffic. The infrastructure traffic carrying capacity can be planned and dimensioned properly. In addition to knowing the TT and BHT data, a SLA can be entered into another carrier for each of the services to be transported.

Leased Line (T1/E1)

Fractional T1/E1 (n*DS0)

Frame Relay (n*64)

   PVC

   SVC

ATM (n*512k) rt-VBR

PRI

ISDN (2B+D)

X.25 (n*64)

The first part of determining the traffic volume to transport involves determining the physical layer dimensioning. After the physical layer has been determined, then the Value Added Services that utilize adjunct platforms can be then determined (that is, e-commerce).

It is important to note that the busy hour for one service type will in all likelihood be different for another service. Specifically, the BHT for PRI traffic will probably be different from that used for IP traffic, if the IP is residential-oriented. To facilitate the discussion for the examples presented, the busy hour is assumed to be the same for each service type. However, in real life, a look at the busy hour requirements for each service type will need to be done because Frame Relay will use a different network than voice services, for example.

The traffic in Table 4-1 needs to be presented in an Mbps level. The data shown in the table is a summary for the system. The engineering community will need to have the data in summary form, as well as geocoded for input into the engineering models for designing the network.

The specific services offered in the table were picked for example purposes and should help structure the thought process of what will or will not be offered. Also, services offered initially do not have to be the same as those offered in later years.

Additionally, the oversubscription levels are not realistic and should be driven by what the market will bear. However, it is in the advantage of the operator to oversubscribe certain services in order to maximize the utilization of the infrastructure. Additionally, if the system does not have dynamic bandwidth allocation available for use, the use of oversubscription is not relative for the radio environment, just the network components.

Table 4-1

Busy hour traffic
(BHT)—full service
LMDS total system
BHT

Grid Number	Type			Number Sub	Oversubscription Factor	Total Mbps
AAL1	**VPN/WAN**	**CIR**	**Burst**			
	T1/E1	1.5M	NA	150	0	225
	T1/E1 (IP)	1M	1.500	500	10	525
	Frame Relay 128k (SPV)	.064	.128	200	10	14.08
	Frame Relay 512k					
	(PVC)	.512	NA	100	0	5.12
	ATM (PVC)	1M	1.5M	25	5	27.5
	X.25	0.064	NA	50	0	3.2
AAL1	**Circuit Switched**					
	BRI (1B+D)	0.064	NA	500	0	32
	BRI (2B+D)	0.064	0.128	250	10	17.6
	PRI (10–24)	0.64	1.536	750	5	710.4
	POTS	0.064	NA	100	0	6.4
AAL5						
	IP (always on)	.256	.512	1000	30	264.53

Also, the Table 4-1 does not differentiate differences in up and down link traffic differences. For example, IP traffic is more downlink, to the terminal, weighted then uplink, from the terminal. That is, of course, unless the customer involved is the server complex itself.

Table 4-1 is meant for a LMDS service provider, which is geared toward the SME and SOHO market. If the service offering is geared toward the residential market and is IP-driven, the following modification shown in Table 4-2 is best done (it assumes that there are not other services offered).

The calculation method is as follows:

$$\text{Total Mbps} = (\text{CIR} + ((\text{Burst} - \text{CIR})/\text{Oversubscription Factor})) \times \text{\# subs} \times \text{\% online}$$

Table 4-2

System busy hour traffic (BHT) IP based total system BHT

Grid Number	Type	CIR	Burst	Number Subs	% online	Oversub-scription Factor	Total Mbps
AAL5							
	IP (Always On) uplink	.128	.512	1000	10	30	14.08
	IP (Always On) downlink	.512	1M	1000	10	30	52.94

Therefore:

$$\text{IP (Always on) Uplink} =$$
$$(0.128 + (0.512 - 0.128)/30) \times 1000 \times 0.1 = 14.08 \text{ Mbps}$$

The impact to the total Mbps, based on the oversubscription and CIR, is evident.

If the system is a duplex system, then the downlink channel for this service is the one that is loaded up. If the system is a TDD system, then the combined uplink and downlink are added together. Of course, the throughput of the radio channel is not 100% efficient and the engineering department will apply an overhead tax, inefficiency factor, to the traffic based on its type.

For example, in the IP example if the system was inefficient in IP transport, the overhead tax may be 25%. In that case:

$$\text{IP Traffic (Downlink)} = 52.94 \text{ Mbps}$$

$$\text{Radio Channel Tax (25\%)}$$

$$\text{Radio Link Traffic} =$$
$$\text{IP Traffic (Downlink)} \times \text{Radio Channel Tax}$$

$$= 52.94 \text{ Mbps} \times 1.25 = \textbf{66.175 Mbps}$$

Therefore, the radio system needs to transport **66.175 Mbps** of data to carry **52.94 Mbps** of real traffic.

The same methodology can be applied to any service support. Therefore, it is equally important to match the service offering to the type of service the LMDS system's infrastructure will best support in the most efficient manner as shown in Table 4-3.

Table 4-3

Total system daily/weekly traffic

Grid Number	Type	CIR	Burst	Number Subs	% online	Oversub- scription Factor	Total Mbps
AAL1	**VPN/WAN**	**CIR**	**Burst**				
	T1/E1	1.5M	NA	150	0	225	
	T1/E1 (IP)	1M	1.500	500	10	525	
	Frame Relay 128k (SPV)	.064	.128	200	10	14.08	
	Frame Relay 512k (PVC)	.512	NA	100	0	5.12	
	ATM (PVC)	1M	1.5M	25	5	27.5	
	X.25	0.064	NA	50	0	3.2	
AAL1	**Circuit Switched**						
	BRI (1B+D)	0.064	NA	500		TBD	
	BRI (2B+D)	0.064	0.128	250		TBD	
	PRI (10–24)	0.64	1.536	750		TBD	
	POTS	0.064	NA	100		TBD	
AAL5							
	IP (Always On) up	.256	.512	1000		TBD	
	IP (Always On) down	.512	1M	1000		TBD	

NOTE: Circuit Switch Traffic can be oversubscribed depending on what part of the infrastructure can make concentration decisions.

When putting together the daily, weekly, and monthly traffic volume, the committed rates for VPN, as shown in the table, do not alter from busy hour to daily, etc.; unless of course the SLAs offered reflect something different. The variable that to the traffic forecasting when looking at the daily rate applies primarily to the circuit switch and IP traffic.

The IP traffic can be selected to have an average throughput with bursting capability to a particular level or have the SLA tailored for usage with a guaranteed bandwidth, CIR, and burst rate. If you know the traffic distribution, the usage-sensitive method may prove to be most cost-effective, again depending on the SLA used.

The data in Table 4-4 include some simple dimension values that can be used to determine the relationship between TT and BHT.

$$\text{BHCA} = (\text{BH Calls/User})/\text{BH Call Attempt Success Ratio} = 6/.75 = 8$$

Most of these numbers can then be input into the following table to arrive at the amount of traffic expected to be transported during the whole day or busy hour. Of course, the table does not account for weekend traffic,

Table 4-4

Standard traffic dimensions

Call Pattern	Value
Daily Erlangs/User	1.2 (72 minutes)
% BH/Total Erlangs	25
BH Calls/User	6
Peak BH Erlang/User	0.12
Compression Ratio	4:1
Line Extension Ratio	
1 line	0.8
2–6 Lines	0.833
7–25 Lines	0.769
26–50 Lines	0.714
51–up Lines	0.666
BH Call Erlangs	0.3
BH Call Duration	3 minutes **
BH Call Attempt Success Ratio	75%
BHCA /user	8

** BH Call Duration will change, based on the type of business/service the customer is involved with.

etc. However, the format can be easily manipulated to meet the desired requirements for any LMDS system.

Another issue related to VoIP and circuit switch traffic, the migration of VOIP, will alter the radio and some of the call treatment. However, for call treatment reasons, the format that is shown should still be used.

Example: This is a SOHO office with seven employees. The distribution in the previous table will be applied to this office.

$$\text{Total Usage} = \text{Number of Employees} \times \text{Line Extension Ratio} \times$$
$$\text{Daily Erlangs/User}$$
$$= (7 \times 0.769) \times (72 \text{ minutes}) = 387.576 \text{ minutes}$$

$$\text{BH Usage} = \text{Number of Employees} \times \text{Line Extension Ratio} \times$$
$$\text{BH Call/User} \times \text{Call Duration}$$
$$= (7) \times (0.769) \times 6 \times 3 =$$
$$96.894 \text{ minutes (ie 25\%)}$$

$$\text{BHCA} = (\text{Number of Employees} \times \text{Line Extension Ratio} \times$$
$$(\text{BH Calls/User})/\text{BH Call Attempt Success Ratio}$$
$$= (7 \times 0.769) \times (6)/0.75 = 43$$

Therefore, this customer has a potential voice usage of 388 minutes per day, with 97 minutes in the busy hour from a total of 43 BHCAs.

Table 4-5 is an example of how the traffic just calculated can be distributed.

NOTE: *The call duration and number of minutes is important when working with interconnect agreements relative to the traffic volume anticipated to be delivered to them. However, for radio-calculation purposes, the number of calls equal to the number of DS0s assigned, based on the line extension ratio, is what is used.*

Continuing with the concept, the traffic distribution for on-net and off-net can be carried to all the service offerings. Again, if a service is not offered, it is nulled out and does not need to be included in the calculations as shown in Table 4-6.

Applying the expected traffic with the agreed-upon distribution for on-net and off-net, the types and sizes of the associated pipes in and out of the LMDS system can be determined.

Table 4-5

Circuit switch traffic

	Incoming	Outgoing	Incoming	Call Duration	Number Calls	Total Traffic (TT)	BHT
(25% TT)							
Total Local Calls	35%	65%				135.8	33.95
On-Net	0%	5%	3 minutes			0	
Off-Net	100%	95%	3 minutes			135.8	33.95
Long Distance	55%	30%	4 minutes			213.4	53.35
International	10%	5%	4 minutes			38.8	9.68

The final step needed is to provide the system data for the type of service, its usage, the amount of on-net and off-net distribution, and its geocode, so that the data can be utilized extensively by the engineering community.

Capital Expenses

Capital expenses (CapEx) are key elements of a budget that is primarily driven by the technical community within an LMDS company. CapEx basically includes all capital that is used in the building of network to support the services sold. CapEx also includes personnel costs that are incurred for the construction and design of the infrastructure. Often, personnel costs when put in budgets, are included solely within the operations expense portion and should have some inclusion in the CapEx budget.

All wireless systems are capital expenditure-intensive. So just what are the items to include for CapEx? The answer is as many items as you can legitimately assign to this category.

The following is a brief list of the items that should be included in CapEx:

Base Station

1. Site Survey

2. Site Acquisition

3. Base Station Design

4. Site Preparation

Table 4-6

Total system BHT distribution

Type	Year 1 On-Net	Year 1 Off-Net	Year 2 On-Net	Year 2 Off-Net	Year X On-Net	Year X Off-Net
VPN/WAN						
T1/E1	0%	100%	10%	90%	30%	70%
T1/E1 (IP)	0%	100%	0%	100%	0%	100%
Frame Relay 128k (SPV)	0%	100%	5%	95%	10%	90%
Frame Relay 512k (PVC)	0%	100%	10%	90%	15%	85%
ATM (PVC)	0%	100%	10%	90%	20%	80%
X.25	0%	100%	0%	100%	0%	100%
Circuit Switched						
BRI (1B+D)	0%	100%	10%	90%	20%	80%
BRI (2B+D)	0%	100%	10%	90%	20%	80%
PRI (10–24)	0%	100%	10%	90%	20%	80%
POTS	0%	100%	10%	90%	20%	80%
IP						
IP (Always On)	0%	100%	0%	100%	0%	100%

5. Land Use Acquisition (Zoning)
6. Site Improvements
7. Site Construction
8. Site Installation
9. Equipment Shelter
10. Antenna Mount/Mast
11. Cabling
12. Cable Tray
13. Installation Material
14. Installation

15. Commissioning

16. Integration

Base Station Equipment

1. Shelves

2. Radio Equipment

3. HVAC

4. Battery Plant

5. Power

6. Primary Reference Source (PRC)

7. Software

Microwave Radio (Point to Point Systems)

1. Radio Equipment

2. Software

3. IDU Radio Equipment

4. ODU Radio Equipment

5. Cabling

6. Antenna Mounting/Mast

7. Installation

8. Commissioning

Central or Node Office

1. Site Survey

2. Site Acquisition

3. Site Preparation

4. Land Use Acquisition (zoning)

5. Site Improvements

6. Site Construction

7. Site Installation

8. Cable Tray

9. Installation Material

10. Installation

11. Class 5 Circuit Switch

Service Modules   dependent upon dimensioning
Switch Software   including right to use fees for anticipated service
  offerings
Installation
Commissioning
Integration

12. Core ATM Switch

    Service Pods
    Software
    Commissioning

13. Edge ATM Switch

    Service Pods
    Software
    Commissioning

14. DXX 3/1 /0

    Software
    Commissioning

15. DXX 3/1

16. DXX 1/0

17. Router

    Service Pods
    Software
    Commissioning

18. VoIP Gateway

    Service Pods
    Software
    Commissioning

19. FaxIP Gateway

    Service Pods
    Software
    Commissioning

20. Surveillance/Wiretap Equipment
21. Primary Reference Source (PRC)
22. STP
23. HVAC
24. IP Server (Proxy/Firewall)
25. Web Server
26. Email Server
27. Transaction Server
28. Network Operation Center
29. Battery Plant
30. Rectifiers
31. Diesel Generator
32. Building Fit-Up
33. Dual Power Routes to Building
34. Dual Facility Routes to Central Office Building
35. Billing System
36. Provisioning System

Host Terminal

1. Host Survey
2. Host Acquisition
3. Host Terminal Location Design
4. Site Preparation
5. Site Improvements
6. Site Construction
7. Host Installation
8. Equipment Cabinet
9. Antenna Mount/Mast
10. Cabling
11. Cable Tray
12. Installation Material
13. Commissioning
14. Integration

Host Terminal Equipment

1. Indoor Unit
2. Outdoor Unit
3. UPS
4. Ancillary Equipment
5. Software

Test Equipment

1. Network Test Equipment
2. Radio Test Equipment
3. Technician Vehicles

The previous list of sources can be easily expanded to include more items and should be tailored to your particular infrastructure vendor, market, and service offering.

Operating Expenses

Because the *operating expenses* (OpEx) for any LMDS system can be high, it is one of the missions of the engineering and operations departments to keep the operating expenses as low a level as possible without sacrificing quality.

Base Station

Lease Expense

Utility Expense (Power)

Facility Expense (PTT)

HVAC Maintenance

Technical Maintenance (2% CapEx)

Yearly Software Fee

Host Terminal

Maintenance (2% CapEx)

Yearly Software Fee

Central Office

Lease Expense

Utility Expense (Power)

Facility Expense (PTT)

Maintenance (2% CapEx)

Yearly Software Fee

Personnel Cost

Management

Staff

Outsourced Staff (Temporary)

There are obviously more items that can be included into the OpEx portion, but the previous list is a simple template from which items can be added.

The facility expenses will be one of the largest aspects (and most difficult to control) of the LMDS system due to the inability to dictate where the customers want to connect the other end of the service circuit or VPN to. Based on the expected volume of traffic (which is based on marketing forecasts), it is possible to obtain volume discounts based on the anticipated volume. There are usually penalties for not meeting the goal, but if it is done right, it is a reasonable risk to take because of the upside savings potential.

Capital Authorization

The authorization to spend capital dollars is an important control item for any wireless system, especially an LMDS system. The requirements for capital authorization vary from organization to organization, but the following fundamental issues should hold.

Prior to installing a central office, base station, or host terminal, their economic viability needs to be reviewed. The review process is rather straightforward and can in fact usually be included with the design review for the base station and central office locations. However, the host terminal requires unique focus.

There are key issues that need to be addressed in any capital authorization for an LMDS system, whether they are of the base station, packet switching platform, or host terminal installation. The following issues need to be identified with a relative degree of specificity. Because there are a plethora of LMDS systems, the commonality between them lies in the following main points:

Cost per host terminal

Cost per base station

Cost per Mbps

Cost per port

Cost to acquire a customer

Cost to deliver a particular service

These is directly tied to the following:

Customers/Radio Channel

Customers/Base Station

Mbps/Customer

Knowing the previous ratios, a guideline can be established that will define the (Investment/Customer), which will then define the (Revenue/Customer).

If the expected revenue per customer for the service offering, base station or host terminal location meets the desired ratio over a specified time period, then the authorization to release capital should proceed.

References

Anthony, Robert, Glenn A. Welsch, and James S. Reese. *Fundamentals of Management Accounting*, 3rd ed. Homewood, IL: Richard Irwin, Inc., 1981.

Anthony, Robert and Vijay Govindarajan. *Management Control Systems*, 9th ed. Homewood, IL: Richard Irwin, Inc., 1997.

Perrault, William D. and E. Jerome McCarthy. *Basic Marketing*, 12th ed. Homewood, IL: Richard Irwin, Inc., 1999.

Smith, Clint and Curt Gervelis. *Cellular System Design and Optimization*. New York, NY: McGraw Hill, 1996.

Smith, Clint. *Practical Cellular and PCS Design*. New York, NY: McGraw-Hill, 1997.

RF Design Guidelines

Introduction

The RF design process used for an LMDS system involves many aspects that draw upon both microwave point to point system designs and cellular/PCS system designs. However, there are unique aspects to an LMDS system, which, although drawing upon other system designs that are well published, utilize a combined or rather hybrid approach.

It is important to note that the initial LMDS system design process is extremely fluid and is equally complex when addressing system growth issues. The design process can be extremely simplistic or elaborate as time and management allow. Because LMDS involves selectively covering areas where the customers and the benefits are stationary, you can reduce significantly the complexity in site location and design.

However, just putting up a single boomer site, while satisfying the initial system requirements for possibly covering the initial customer or customers, may in fact limit the overall throughput or capacity-carrying ability of the system. For instance, if the site is too high, it may limit frequency reuse. In addition, reliance on a single site for a geographic area leads to simplification of the design, but can restrict new customers from gaining access to the system due to shadowing effects.

Included with the design process are the edge terminal location issues. Primarily, the system design needs to be done so that the alterations to edge terminal locations, where the customers are actually connected to the LMDS system, be minimized. The reason for this concern lies in the physical logistics of potentially having to alter the orientation of all the edge terminal devices when new sites are added, sectors are split, or cell splitting takes place. Although the edge terminal alteration issue cannot be completely resolved in every case, care must be taken in the initial design to ensure that this problem is minimized as much as practically possible.

What follows is a discussion of the design process, the particular nuances that are relative to LMDS systems, and how capacity assumptions play into the design approach. There are, of course, the critical issues of what frequency band(s) the system will operate in and what regulatory requirements and restrictions have been placed upon the system's operating conditions.

It can never be overemphasized in the initial design phase that the amount of traffic expected to be handled by any one site or even the system in total should be reviewed prior to implementation. The design review stage is the best point in the design process to screen potential site problems. The initial capital authorization for the system itself should verify that the capacity envisioned for the system, using the expected services offered with the appropriate tariffs applied, should prove the economic viability of the system. For example, if a site is needed for handling only a total of 256kbps of data and voice traffic combined, the solution for that geographic location may be reselling landline service or abandonment. A similar analysis can be carried forward at the system level.

With that said, the primary question that any design engineer is faced with when beginning the design is determining where to begin. In order to begin the design, the fundamental concept of what the purposes, or criteria, for the design need to be established. Then, there is the issue of what will be the output of the project, when this information will be released, and how many steps it will take.

This chapter will therefore go over the system design requirements for putting together or expanding an existing LMDS system. Topics included here involve the process that an engineer should utilize for designing a broadband LMDS system from the ground up or expanding an existing system. Issues that are covered involve what information is needed to effectively design a network and some proposed output forms. There are many technology and market-specific issues that need to be addressed when putting together an RF design for an LMDS system.

As with any wireless system, the RF system design planning is exceptionally critical. In the RF design, specific technology issues, both current and future, must be factored into the design. If the RF system design does not attempt to address current and future configurations, the potential is there for the system to not meet the market requirements and thus become more costly in the long-term, or worse, cease to be competitive in either cost, functionality, or both.

Additionally, the RF design, covered in the next chapter, must be dovetailed into the network system design to ensure that the network backbone is in place to support the RF design and that the RF design supports the network requirements. The network and RF design process are intertwined and dependent on each other. Therefore, it is important to ensure that a holistic approach is used instead of a compartmentalization or departmentalization in any design process to ensure that a proper design is completed.

RF System Design Process

The RF system design process for an LMDS network is an ongoing process of refinements and adjustments that is based on a multitude of variables, most of which are not under the control of the engineering department. However, the RF design for a new system or the growth analysis for an existing system is meant to provide a road map or plan from which the limited resources—in terms of capital, personnel and company time—can be effectively directed and utilized.

- System Configuration
- Alterations to Existing or Planned System Configuration Over Design Period
- Capacity of System (Sites) in Mbps
- Location of Radio Sites
- *Infrastructure Requirements*   Radios
- Frequency Plan
- Activation and/or Configuration Change Dates

The major output of the RF system design is the required on-air dates for new sites or system alterations or simply channel adds with an existing system because they define when the cell site or change is specifically needed. Obviously, the on-air date for a new system is the on-air date for the entire network, and the sequence that they secured is not as important as that of ensuring that they will all be available for system activation.

The other aspect of the RF system design is to factor the amount of physical radios or rather traffic-handling capacity for each cell site and sector capacity into the design process for an LMDS system. It is referred to as Mbps, with the design being a Mbps/km². As of this writing, there are no effective computer modeling systems for helping define the required channel elements for an LMDS cell site for both downlink and uplink, although several are in the conceptual stage. The process presented here can be used for a TDD, FDD, or any combination.

The process that should be followed is as follows:

1. Determine technical requirements
2. Establish or determine spectrum available for potential use
3. Determine design time frames
4. Establish marketing requirements
5. Determine implementation methodology
6. Make technology decisions
7. Define the types of cell sites (4- or 8-sector)
8. Establish a link budget
9. Establish a PMP grid system
10. Define coverage requirements

11. Define capacity requirements
12. Complete RF system design .
13. Issue search area
14. Site acceptance/site rejection
15. Land use entitlement process
16. Integration
17. Hand over to Operations

Obviously, there are many steps that are incorporated into each of these process milestones. But the process outlined shows the basic process steps that need to be followed by the LMDS RF Engineering department to design a system, regardless of the technology platform chosen.

At each step of the process, a brief design review should take place to ensure that the design is being put together in accordance with the requirements for the network. If the design does not meet the market requirements, the potential for failure is very high.

Methodology

The methodology for the RF system design needs to be established at the beginning of the design process. The establishment of the methodology utilized for the formulation of the design, and ultimately the RF system design report, is essential in the beginning stages to ensure that the proper baseline assumptions are used and to prevent labor-intensive reworks and teeth gnashing.

Some of the methodology issues that need to be identified at the beginning of the study are as follows:

- Time frames for the study to be based on
- Subscriber demographic information (geographic locations where they are clustered)
- Services offered, both future and legacy issues
- Subscriber usage projection (current and forecasted by quarter) by service type
- Circuit-switch migration to IP usage forecasting
- Design criteria (technology-specific issues)

- Baseline system numbers for building on the growth study
- Cell site construction expectations (ideal and with land use entitlement issues factored in)
- New technology deployment and time frames
- Budget constraints
- Due date for report

The time frames for the report are essential to establish prior to the beginning of the system design; this is the report generation. The time frames will define what the baseline, foundation, and how much of a future look is presented in the report. For a new system, the on-air date obviously will be the starting point. However, this is only one milestone because this is when the system loading will officially begin. The design time frame needs to account for how long the design will be valid after system turn-on, assuming that the marketing input and construction assumptions remain valid.

If the LMDS system already exists, the baseline data used will need to be decided upon. The baseline time frame is usually a particular month and the design time frame that is selected will determine which set of data will be used to generate the report. The baseline is critical for an existing system because if the wrong month is selected, the data used for generating the design will alter the outcome of the report by either overprojecting or underprojecting the amount of sites required.

Another nuance with LMDS systems for traffic projections addresses the data and circuit switch service offerings and their time-of-day usage alterations. For instance, time of day is important to define when offering service to both business and residential customers. Residential customers could easily be part of a high-rise or smart building; not just associated with individual homes. Business IP traffic may be more centered at two or three peak times during business hours, whereas residential IP traffic may be focused more in the early evening. This differentiation of service times is important for ensuring that capacity requirements are met and also to potentially foster oversubscription of capacity in any given area.

The traffic projections will also be dependent upon the type of services offered. For instance, if the CBR traffic is offered for leased line replacement, the part of the spectrum that is assigned for this particular service will need to be eliminated for possible other users to share bandwidth with. The obvious implication of this concept of fixed bandwidth allocation is a misapplication of LMDS because it is really acting as a point to point link. However, there will almost always be some level of CBR traffic in the network—particularly leased line replacement—due to the need to generate cash flow in the initial system's life cycle (provided that the infrastructure can support CBR).

Therefore, the amount of time the report projection is to take into account and the services offered are critical for the analysis. The decision to project one year, two years, five years, or even 10 years has a dramatic effect on the final outcome. In addition to the projection time frame, it is important to establish the granularity of the reporting period: monthly, quarterly, semi-annually, annually, or some perturbation of them all.

To properly project the capacity-handling requirements of a system, marketing plans need to be factored into the design. The marketing department's plans are the leading element in any RF system design. The basic input parameters needed for the RF design from the marketing department are listed next. The information is critical, regardless of whether the system is a new or current system.

- Identification of key coverage areas in the network for system turn-on

- Projected subscriber growth for the system over the time frame for the study

- Projected CBR and UBR traffic, and types of service offerings per quarter

- Demographic breakdown of customer types, services, and their logical locations

- IP and circuit-switch traffic blends with potential migration percentage of existing customers

- Edge terminal location multiplication factor (that is, number of customers per edge terminal)

- Special promotional plans over the study time period (such as all-you-can-eat Internet at 512kbps)

- Top 10 potential sales areas for the system (should match demographic data provided)

There is a multitude of other items needed from the marketing department for determining RF system design. However, if you obtain the information from the marketing department on the basic topics listed here, it will be enough to adequately start and complete the RF system design.

The establishment of the design criteria used for the report is another key element when putting together the plan. It is recommended that the design criteria used for the study be signed off by the Director of Engineering to ensure that nothing is missed.

The items in the RF design criteria need to include the following as a minimum:

- Marketing input
- RF spectrum available for the RF plan
- Type of services that will be supported by the RF plan
- Coverage requirements
- Identification of coverage sites
- BH peak traffic; 10-day high average for month (existing system), which may involve day- and nighttime patterns
- Infrastructure equipment constraints
- New technology considerations
- New cell site configurations
- Frequency reuse
- System alterations that are available for the design engineer to utilize, (for example, radio additions, polarization changes, sector splitting, or cell splitting)
- Baseline system numbers
- Cell site deployment considerations

This list should be used as the foundation that can be built upon for establishing the design criteria for the RF system design plan.

The baseline system numbers should be listed in the design criteria for the growth plan. The baseline numbers involve defining the time period for basing the growth projection on. In particular, the decision to use June data instead of July data, for example, has a significant impact in the plan as well as using a peak versus an average value for traffic.

A critical issue with LMDS and the design criteria falls into the traffic assumptions used, coupled with the blend of services offered. The use of CBR and UBR traffic is important to distinguish how the system addresses access by the edge terminal device(s). For example, a design criterion may be to oversubscribe IP data by a factor of 30, which is directly dependent upon the SLA that is offered with this service.

The decision of how much is on-net or off-net for each service has a large impact on the design aspect. For instance, on-net traffic removes the potential for a CLEC to deliver part of the service, but it also increases the traffic load on the network. Specifically, the traffic load is now doubled for the system because the originating point and terminating point both use the same amount of bandwidth, but you can only charge for one. Therefore, you need to factor in potentially additional edge terminal equipment and the additional traffic capacity-handling requirements for the recipient of on-net traffic.

The construction aspects of the system will also need to be factored into the report itself. The construction aspects pertain to the proposed new cell sites that are or will shortly be under construction. Other construction aspects involve the possibility of actually building what is requested in the specified time frame dictated in the report. The realistic time frame for construction is the reality dose that needs to be placed into the report, saying that although 100 sites are needed in the next three months, only 20 will actually be available for operation.

The construction aspects of the design may force a redesign to take place to accommodate the RF coverage or capacity requirements. Specifically, if a site is needed for coverage in a area within three months but the land use entitlement process may take upwards of 12 months, an alternative design will need to be sought. The alternative design may be where the landline service is resold from a CLEC until a long-term solution is available, if ever.

New technology deployments will also need to be factored into the plan. New technology deployment might be converting from a TDD system to a FDD system, or a change in the edge solution equipment utilized. Other technology deployment issues that need to be factored in might be the cell site infrastructure equipment where a new vendor's equipment is planned to be deployed. Also involved with technology deployments could be the introduction of a new modulation formats available for enhanced capacity handling.

The budget constraints imposed on any operator must be folded into the RF plan. Failure to incorporate budget constraints into the plan will only make the plan unrealistic and squander everyone's time needlessly. When putting together the report, any budgetary constraints that exist must be incorporated into the plan. Often, if the plan calls for more sites than budgeted for, then it is imperative to help establish a ranking order for which sites are truly needed versus which are nice to have.

The last part of the preparation for the report is finding out when the report itself is due, the level of detail required, and who is assigned to work on the report itself. All too often, a plan will be put together, but for various reasons the actual due date is often allowed to slip. The negative impact of this ties directly to providing some guidance to the other departments that need the RF design to complete their work because the RF design is usually the gating project for their efforts. If the RF design is taking longer than planned, it is advisable that an interim report be issued to allow various departments, primarily real estate, to begin their efforts.

Technology Decision

The technology decision is a critical issue when putting together an LMDS RF design. Specifically, the issues associated with an TDD system are different from those for a FDD system. The technology chosen will have a profound impact on how the system will be designed.

For instance, TDD systems are designed to handle a non-symmetrical traffic between the upstream and downstream traffic, and FDD is inclined to handle symmetrical traffic better. But TDD has some unique interference concerns and may be easier to deploy for an initial system in a congested RF environment. The FDD system may have more traffic-handling capacity for the same spectrum available due to C/I considerations. But at this time, neither the system type or hybrid variant has been deployed in any significant number to determine under exactly which condition one type versus the other would make the most logical choice.

What is more likely is the decision is to utilize multiple technology platforms within the network for a given market. The obvious concern and issue deals with the various compatibility issues that need to be factored into the design process. For example, if a hybrid band approach is used where 3.5GHz is used in conjunction with 26GHz, the range and capacity-handling capabilities of each will need to be brought into the design process for determining the location and the percentage of blend that will be used.

The choice of bandwidth enhancements needs to be brought into the design process. Depending on the type of service offered, the use of bandwidth enhancements can be effectively used. For instance, the main bandwidth enhancement is *Dynamic Bandwidth Allocation* (DBA). The use of DBA will allow for oversubscription and reduce the capital cost for infrastructure.

Another area to consider for technology decisions lies in the edge terminal solution. For instance, if the edge terminal can only support IP thus if the decision is to offer POTS as part of the product portfolio, it will present a potential problem. Obviously, there is the VoIP solution using an adjunct platform, but unless the latency and throughput requirements are met (to mention a few critical issues), then the delivery of this service is in question. Alternatively, if the edge terminal offers POTS, ISDN (BRI), and other related interfaces for services but the marketing plan is to offer IP traffic only, then the edge terminal solution chosen needs to be readdressed.

The obvious point with the technology decision made will not only impact the services that may be provided by the operating company to the subscriber, but also have a profound impact on how the network is laid out.

Link Budget

The link budget used for an LMDS system's design is one of the key technical parameters used for the design process. Obviously, all the components in the design process are important, but the link budget directly determines the range and deployment pattern used for a given LMDS system. An important aspect is that there can be several link budgets in any system, based on the operating frequency, spectrum allocated, link reliability, physical components for the system, differences in up- and downstream modulation methods or protocols, and rain fade issues (to mention some of the elements that need to be considered).

Additionally, the link budget decided upon for the system design needs to account for not only the physical issues defined, but also marketing issues. Marketing issues address the path reliability deemed necessary for the market that is being targeted. For instance, if marketing determines that a link reliability of 99.99% is needed, then adhering to 99.999% as a design makes little sense. Conversely, if a goal of 99.999% is desired for the service offering, then designing the network for a 99.99% can result in a significant problem. The overdesigning of the network has advantages in terms of link reliability in that it builds in the added buffer for marketing changes in the future, based on outside competition requiring greater reliability. However, in the design process, when presented by marketing a reliability number, the radio link is only part of the reliability factor and this fundamental issue needs to be accounted for in the totality of the system design process.

The objective behind the link budget is to determine the path length or rather the size of the cell sites needed for the network design. Specifically, the D/R ratio, in addition to the link reliability that is chosen for the system, will determine the radius of the site and also the distance between the LMDS cells for coverage and reuse considerations.

A sample link budget is included in Tables 5-1 and 5-2 for reference. The link budget is meant to accommodate multiple technology platforms and transverse over a wide range of radio spectrum issues. The link budget format presented here can and should be used when putting together the

Table 5-1

Downlink link
budget

Downlink		Units
Base Station Tx PA		dBm
PA Backoff	( )	dB
Tx Filter	( )	dB
Combining Loss	( )	dB
Cable and Connector Loss	( )	dB
Antenna Gain		dBi
Attack Angle	( )	dB
Fade Margin (99.999%)	( )	dB
Adjacent Cell Overlap	( )	dB
Edge/Host Terminal Rx Antenna Gain		dBi
Noise Figure (NF + Cable Loss)	( )	dB
C/N	( )	dB
Rx Sensitivity (x BER)		dBm
Downlink Path Gain	XX	dB

Table 5-2

Uplink link budget

UpLink		Units
Edge/Host Terminal Tx PA		dBm
PA Backoff	( )	dB
Tx Filter	( )	dB
Cable and Connector Loss	( )	dB
Antenna Gain		dBi
Fade Margin (99.999%)	( )	dB
Base Station Rx Antenna Gain		dBi
Noise Figure (NF + Cable Loss)	( )	dB
C/N	( )	dB
Rx Sensitivity (x BER)		dBm
Uplink Path Gain	XX	dB

design of a system. There are some issues that are covered in the link budget that may or may not pertain to the particular potential cell site or technology platform utilized, such as rain attenuation, when operating in the 2.5GHz or 3.5GHz bands.

The tables capture most, if not all, of the important gain and loss components that comprise the link budget. The reason for conducting both uplink and downlink calculations is to establish the weakest link in the design path and from there to set the cell radio requirements. In many LMDS systems, the path does not require being balanced, only that the path gain, with the appropriate link reliability value, is equal to or greater than the path loss determined.

The attack angle value, normally not shown in link budgets, allows the system design engineer to enter a value to derate the antenna gain to account for system installation variations. Specifically, what this refers to is the issue that only a few host terminals will be installed on-site to the base station's sector antenna, thereby enjoying the maximum gain obtainable. Therefore, if using a 90-degree sector, the antenna gain is 3dB down at the sector borders, and therefore this is the attack angle value that could be entered. The attack angle is only factored into the link budget portion associated with the base station because it is assumed that the host terminal's alignment is optimized.

The following relationships have to hold for the LMDS system to perform:

Downlink Path Gain $\geq$ Pathloss for LOS

Uplink Path Gain $\geq$ Pathloss for LOS

Provided that both of these conditions are met, the decision of which path is the limiting case needs to be established by using the following method:

Downlink Path Gain $-$ Uplink Path Gain $= X$ (dB)

Uplink Path Gain $-$ Downlink Path Gain $= Y$ (dB)

You are to use the limiting case for the design that will be either the uplink or downlink path for an unbalanced system. In a balanced system, the choice is irrelevant. However, in the previous relationship, if Y is less than X, the uplink path is the limiting design case.

Table 5-3 is a simple link budget example. The assumptions used for the link budget are also discussed. The values listed are for example purposes only and should not be used for a design because each transport technology is different (in addition to the vendor-specific equipment parameters).

The site is a sector system that uses a Tx PA of 5 watts (37dBm). The base station Tx and Rx uses a duplex antenna system where the PA and Rx is integrated into the antenna itself. The nominal antenna gain is 27 dBi.

Table 5-3

Link budget

UpLink		Units
Edge/Host Terminal Tx PA	37	dBm
PA Backoff	(3)	dB
Tx Filter	(0.5)	dB
Cable and Connector Loss	(0)	dB
Antenna Gain	21	dBi
Fade Margin (99.999%)	(20)	dB
Base Station Rx Antenna Gain	27	dBi
Noise Figure (NF + Cable Loss)	(4.5)	dB
C/N	(15)	dB
Rx Sensitivity (x BER)	−85	dB
Uplink Path Gain	127	dB

The PA backoff for stability purposes is listed as 3 dB, meaning that we really Tx at 2.5 watts. For implementation reasons, described later, a cell overlap was picked as 3 dB. The sensitivity of the receiver for the base station was picked at −85 dBm, but will of course change based on the bandwidth of the radio as well as the modulation format and required BER.

The edge/host terminal parameters were similar to the base station in that the sensitivity of the receiver was again −80 dBm, but the nominal antenna gain was picked at 23 dBm. A PA backoff for the site was also chosen as 3 dB. The PA and receiver are integral to the antenna itself but have a noise figure of 4.5 dB.

For both the base station and host terminal, the fade margin is chosen as 20 dB. The fade margin is established by reading the ITU chart based on the operating frequency and the rain zone that the system operates within (20 dB was arbitrarily chosen for the example). Additionally, the C/N was decided to be 15 dB for this example. Again, the required C/N is dependent upon the bandwidth, modulation, and acceptable BER for the system.

Downlink Path Gain > Uplink Path Gain

In this example, the uplink path is the limiting case for the design for establishing the maximum path for the site.

An important aspect in the previous example is that modulation schemes were assumed to be the same for both base and host terminals Tx, but in many instances for LMDS the modulation format may be different depending on the type of payload services the system is meant to deliver.

Path Loss Calculations

The method for determining path loss for an LMDS system is both simple and complex, allowing for the gray non-answer that is so often encountered when receiving a technical answer to an apparently simple question. For systems normally referred to as LMDS 26, 26, 28, and 39 GHz systems, the path calculations are purely line of site. While the bands for MDS, WCS, and UNII rely on line of site also, but MDS and UNII also can tolerate in some obstructions (this comes at the expense of the link budget and thus the range for the cell).

The free space path loss calculation is the following:

$$PL \text{ (free space)} =$$
$$32.4 + 20\log(R) + 20\log(f)$$
$$\text{where } R = \text{km and } f = \text{MHz } [1]$$

The following calculation is a simple comparison of differences in range, based on the frequency of operations that a system may have. The path gain used for calculations in Table 5-4 was the value used in the link budget example for uplink.

$$\text{Range of Site/Sector} =$$
$$\text{anti-log } ((32.4 - 20\log(f) + \text{Path Gain})/20)$$

Where path gain = 127 dB. (from the previous link budget calculations)

Table 5-4

Base station radius versus frequency

Frequency (GHz) $32.4+20\log(f)$	Frequency Attenuation (dB)		
	Path Attenuation (dB)		Maximum Radius (km)
2.6	100.69	26.31	20.67
3.5	103.28	23.72	15.34
10.5	112.82	7.18	2.28
26	120.69	6.31	2.067
39	124.22	2.78	1.377

In practice at 5km, the attenuation slope used for path loss increases to 35 db/decade, but this example shows the impact that the operating frequency has on the reach of the site and also the relationship with the link budget.

Additionally, the particular radio platform chosen will have a direct impact upon the both the transmit and receive characteristics. A simple example for the impact that bandwidth has on the link budget is shown as follows. This example revisits the bandwidth issue of comparing a 7MHz channel to that of a 14Mhz channel. In the previous link budget example, a receiver sensitivity was defined as −85 dBm for the required BER. For simplicity issues, the sensitivity value is associated with the use of the 7MHz channel. Therefore, the 14Mhz channel is by default less sensitive than the 7Mhz channel, assuming the same modulation scheme.

$$\text{Gross Sensitivity Change (dB)} =$$
$$10 \log (\text{Radio A Bandwidth} - \text{Radio B Bandwidth})$$

Assigning the 7Mhz radio as A and the 14MHz radio as B, the equation can then be worked out:

$$\text{Gross Sensitivity Change (dB)} = 10 \log (7\text{Mhz}) - 10 \log (14\text{MHz})$$
$$= 10 \log (7 \times 10^6) - 10 \log (14 \times 10^6)$$
$$= 68.45 - 71.46$$
$$= -3.01\text{dB}$$

Therefore, by doubling the bandwidth of the receiver, the sensitivity was decreased by 3 dB (in this case, to −82 dBm from −85 dBm), resulting in further reducing the range of the site's or host terminal's reach.

LMDS Cell Sites

The type of cell site(s) that are chosen for the design of the network will have a profound impact on the overall success or failure of the network design. For instance, a desired goal is to design a low recurring cost system that is not capital-intensive for the LMDS operating company. The number of cell sites is needed to be factored into the design, in addition to their impact on recurring costs. The recurring costs of the system involve leases, facilities, and other maintenance issues, which also include engineering efforts.

The use of smaller cell sites is one way of reducing the lease costs for a system. However, if this is used for an initial system deployment, the sheer number of cell sites required to cover the same geographic area will increase. On the other hand, if taller sites were chosen as the design of choice, they would need to potentially be larger in physical size in both equipment and antenna real estate. There is, of course, a blend or middle ground that could be used with large and small cell site designs to potentially meet the requirements for the network.

The different types of cell sites are included next to illustrate the number of different types of cell sites that can simultaneously exist in a network. The proposed listing that follows is not all-inclusive and there are many perturbations to those shown next.

1. 2-Sector (90,0,90,0-degree sectors)

2. 4-Sector (90-degree sectors)

3. 6-Sector (60-degree sectors)

4. 8-Sector (45-degree sectors)

5. 12-sector (30-degree sectors)

The first type of configuration is a 2-sector site shown in Figure 5-1. This type of site uses two or possibly only one sector for immediate operation because that is all that is required for coverage and capacity purposes. The other sectors are not operational at this time, but could be made operational

Figure 5-1
Two-sector LMDS
base site

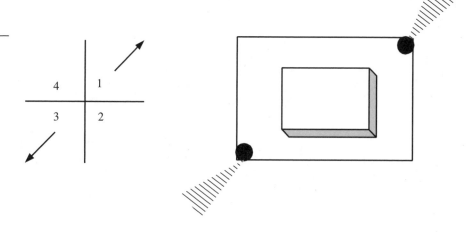

with minimal effort, provided that the coax cables and appropriate mounting brackets are preinstalled.

The next type of site represented is a continuation of the previous site, except all four sectors are populated with radio equipment. The diagram could easily represent multiple channels per sector being used and the drawing assumes that the combining process takes place in the equipment shelter, minimizing the amount of physical antennas, cables, and other lease cost enhancers. The 4-sector-type site, shown in Figure 5-2, will most likely be the predominant type of LMDS site deployed for capacity and coverage combinations.

The next site configuration shown is the 6-sector-site design shown in Figure 5-3. This type of site design has higher density, Mbps/km², than a 2-

Figure 5-2
Four-sector LMDS base site

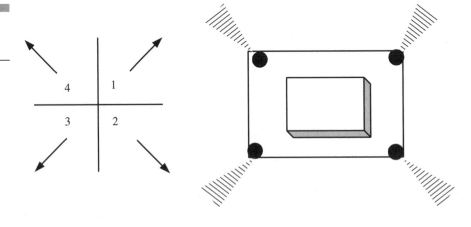

Figure 5-3
Six-sector LMDS base site

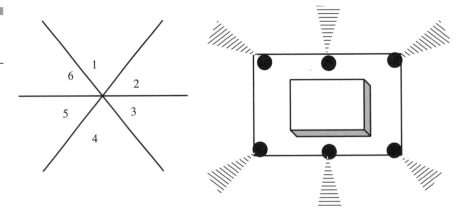

Figure 5-4
Eight-sector LMDS
base site

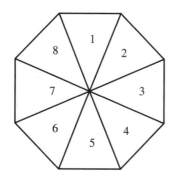
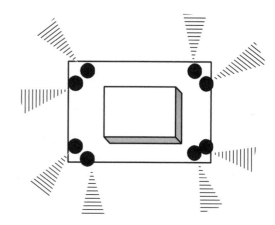

Figure 5-5
Twelve-sector LMDS
base site

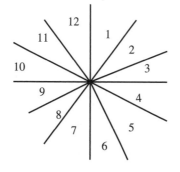
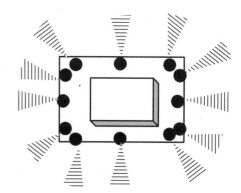

or 4-sector system. This design is used in systems that are allocated multiple of three frequencies. The use of 60-degree antennas can also be used to enhance the capacity of a 4-sector LMDS system by replacing three of the adjoining sectors with three 60-degree sectors.

The site in Figure 5-4 is an 8-sector site that is used when capacity is needed for a given area, and uses 45-degree sectors. The increase in sectors enables a higher density of Mbps/km² to take place.

The fourth example shown in Figure 5-5 of an LMDS site involves a 12-sector site, which of course uses 30-degree sectors and provides the highest density of Mbps out of the LMDS configurations presented.

It must be stressed that mixing different LMDS site configurations, especially 90 degree with 60 and/or 45 with 30 and so on, has design implications associated with frequency reuse and needs to be seriously weighed and reviewed prior to implementing.

For all the configurations shown here, it is assumed that each radio uses the same amount of radio spectrum (that is, has the same channel bandwidth). Again, if this is not the case due to capacity or a vendor discontinuing a product line, the frequency reuse considerations need to be heavily evaluated for potentially negative consequences.

Utilizing the link budget, a rough estimate of the number of cell sites that are required for network design can be established. The rough estimate for cell sites needs to factor into capacity requirements for the network. At this stage of the design process, the particular offloading adjustments for each cell site do not need to be accounted for. This is where an estimate of the type and amount of cells needed for the design is identified.

If the initial cell site count is above what the business case indicates, an interactive step needs to take place, where the assumptions used for the initial cell site count need to be adjusted and recalculated.

Coverage Requirements

One important step in the RF system design process is to determine all the coverage requirements needed for the network. The coverage requirements for a network will obviously be different for a new system from that of an existing one. That is, you are trying to provide service to an area where your marketing and sales department believes they have the best chance for obtaining initial customers. For an existing system, the coverage requirements are designed to expand the sales potential for designated areas.

It must be stressed that LMDS system coverage is fundamentally different from that of a typical wireless system in that coverage is not ubiquitous for the market, but highly focused to areas where only existing or potential customers are located. LMDS service is a fixed service and therefore does not require a contiguous network in order to offer service. But it is advisable that more than one site be deployed in any market to ensure complete coverage of potential customers' locations within a very defined market area. As implied by the previous comment, arriving at coverage requirements for a new or existing system is not straightforward and requires careful planning.

Now comes the Catch-22 part of LMDS designs. An LMDS system allows selective targeting of customers through deploying sites that are meant to cover only customers that will need the services provided. However, you can not hook up any customers, until there is some level of coverage to their facility, because it is wireless. So, you have to build a few sites in anticipation of obtaining customers. If the area that you want to serve is the business center of a city or a residential market, the coverage definitions are fundamentally different in obtaining structures suitable for the installation. Regardless, the coverage analysis must take place using the following general process.

The process defined can be used for an initial system build or the expansion of an existing system.

RF Coverage Identification Process for a New System

1. Marketing and Sales define coverage requirements.

2. Use the design criteria used by RF Engineering to determine how many sites will be needed to satisfy the design goals.

3. Use the list of cell sites identified in Step 2, and rank order them according to the point system methodology.

RF Coverage Identification Process for Existing System

1. Coverage requirements are identified by the following:

   Marketing and sales
   System performance: capacity constraints, link reliability, shadowing, etc.
   Operations: access, security
   Customer care
   RF Engineering: capacity or frequency planning

2. Generate a propagation plot of the system or sub-regions that reflects the current system.

3. Generate a propagation plot of the system or sub-regions that reflects the current system and the known future cell sites under construction.

4. Using plots from parts 2 and 3, compare them against areas identified in part 1 for correlation.

5. Using the design criteria used by RF Engineering, determine how many sites will be needed to satisfy the design goals.

6. Using the list of cell sites identified in Step 6, rank order them according to the following point system methodology.

The point system will involve four key parameters—all ranked from a scale of 1 to 5—based on their severity. Each of the five categories receives a value that is then multiplied by each field to arrive at a ranking. The ranking methodology uses the following key fields:

- System Performance Requirements
- Capacity Relief (Mbps)
- Marketing and Sales Needs
- RF Design

To help clarify the ranking methodology, Table 5-5 has values added to assist in the explanation.

This simple example shows that for an existing system, cell 102 has a higher weight then 101 in the identification of prioritization. The ranking methodology should be applied to all the potential cell sites in a network. The rationale behind establishing a ranking system folds into the budgetary issues of having a method in place to determine where to cut or add sites to the build program.

The following example involves determining the approximate amount of cells needed for covering a specific area of the market that has been defined as attractive to provide service to.

Example It is desired to know how many sites may be required to provide coverage for the core of a market using 39GHz spectrum. The amount of cells required can then be used to determine the initial capital outlay required (there are, of course, other components to this).

Cell radius (from link budget) = 1.37 km

Area of a cell = 5.89km^2

Total area to be covered = 100km^2

$$\begin{aligned}
\text{Number of LMDS Base Stations Required} &= \text{Total Area/Area of Cell} \\
&= 100/5.89 \\
&= 16.97 = 17 \text{ Base Stations}
\end{aligned}$$

Using the 3.5 GHz band, the following is the result:

Table 5-5

Ranking

Cell	System Performance	Capacity	Marketing/ Sales	RF Design	Total
101	2	5	1	1	10
102	1	3	5	2	30

Cell radius (from link budget) = 15.34 km

Area of a cell = 737.34 km²

Total area to be covered = 100km²

Number of LMDS Base Stations Required = Total Area/Area of Cell

= 100/737.34

= 0.13 = 1 Base Station

The obvious issue is that for just providing coverage, the lower band will provide the coverage easily for the target area with one site versus 17 given the same parameters. The previous example illustrates that a dual band approach for obtaining customers may be very attractive. Specifically, the lower band can be deployed in an area by using licensed or unlicensed spectrum. Once the customers are obtained, then the decision can be made about where to place the base stations, thereby reducing the overall capital required for deployment using the 24GHz, 26GHz, 28GHz, or 39GHz bands.

To further complicate coverage issues, it has been found that the deployment of multiple base stations is advantageous for providing coverage for the target area due to the variability of building heights at which the edge terminals will be installed. Specifically, the LMDS base station typically is located on one of the higher structures for a given area, as shown in Figure 5-6.

As shown in the Figure 5-7, although *Line of site* (LOS) exists for many of the buildings for the target area desired, there are some buildings in which LOS is not achievable.

Figure 5-8 illustrates problems that may occur due to a potential customer being located in a building adjacent to the LMDS base station.

Figure 5-6
LOS

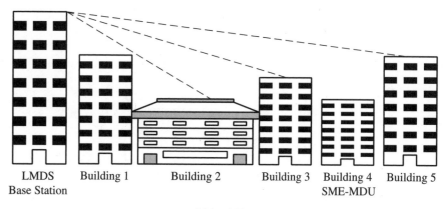

LMDS Base Station   Building 1   Building 2   Building 3   Building 4 SME-MDU   Building 5

Line of Site

Figure 5-7
Non-LOS

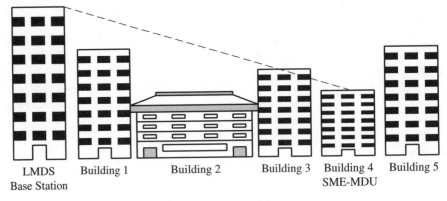

LMDS   Building 1   Building 2   Building 3   Building 4   Building 5
Base Station   SME-MDU

Non-Line of Site

Figure 5-8
Proximity non-LOS

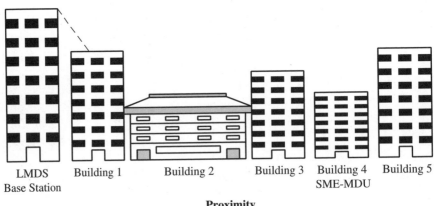

LMDS   Building 1   Building 2   Building 3   Building 4   Building 5
Base Station   SME-MDU

**Proximity
Non-Line of Site**

Therefore, to overcome many of these simple installation problems it is advisable that there be sufficient LMDS base station overlap to ensure that the variability in building heights is resolved and the maximum sales potential can be recognized. The use of a three-base-station-system deployment as a minimum for any geographic area is recommended, provided that the objective is to serve the core business district of a market, for example. See Figure 5-9.

Figure 5-9
LMDS triad

LMDS HUB #1

LMDS HUB #2

City

LMDS HUB #3

Capacity Cell Sites Required

The next step in the RF system design is to determine the LMDS sites required for capacity relief, which can be either for a new or existing system. Obviously, if the system is new, the potential for capacity relief is minimal. It would be a pleasant surprise if the realistic demand for the service would put the system into capacity constraints. If the LMDS system is used for leased line replacement only, however, capacity constraints could become a problem with a new system. The process shown for determining capacity requirements can easily be used in any of the LMDS technology platforms deployed.

If you are looking for a single process that fits all situations, you may have to read tea leaves because the capacity calculations are estimates of take rates of various services that are sold. The capacity calculations are further complicated by service offerings that involve a host of SLAs and the inclusion of CBR, UBR, rt-VBR, and non-rt-VBR. But, as with all estimates, good solid rules can be applied from which reasonable estimates can be arrived at for the capacity requirements.

It is strongly recommended that the amount of traffic expected to be handled by any one site or even the system in total is reviewed prior to implementation. The design review stage is the best time to screen potential site problems, while the initial capital authorization for the system itself should verify that the capacity envisioned for the system, using the expected services offered with the appropriate tariffs applied, should prove the economic viability of the system. For example, if a site is needed only for handling a total of 256kbps of data and voice traffic combined, the solution for that geographic location may be reselling landline service or abandonment. A similar analysis can be carried forward at the system level.

The first step in any capacity analysis is to review the design criteria and evaluate the data that is available. You need to define what it is exactly you are intending to provide for service. For instance, is the service offering leased line replacement only, IP traffic only, dial tone service, data services such ad Frame Relay, or a combination of all, plus some? It is critical to define what you intend to sell now and what you may sell in the future (that is, when you grow up), but either way the choice of service types and their offering levels has a direct impact on the system planning and ongoing support. It is also understood that market conditions, as they are meant to do, will cause shifts in service offerings and those changes should be reviewed in order to understand their impact on the system and the ability to meet whatever the SLA and take rates that the change represents.

In both new or existing systems, all the data necessary to perform a complete design is often not available or is not made available in a timely fashion, creating a large amount of stress upon the design team. Once all the available data is collected, this information should be compared against what was initially desired to have obtained; then, make the necessary adjustments and guesses, based on the missing material. It is rare when any design process has at its disposal all the required data.

The first step in the process involves determining which sites and/or sectors expend their current capacity and require some level of relief. The time frame for capacity exhaustion is very important in that you do not want to deploy equipment or cell sites too much in advance of capacity exhaustion. Deployed too late, however, it would have potentially negative customer satisfaction impacts.

The capacity relief can come from a variety of options, but the most likely possibilities are listed as follows in ranked order:

1. Radio additions

2. New sites

3. Redirecting specific edge terminals to sites with excess capacity

4. Modulation format alterations

5. Antenna system alterations (that is, sector splitting)

6. Service level adjustments

7. Offloading to point to point microwave system or landline facilities

8. Price point changes from marketing (that is, increasing service price to remove its attractiveness)

9. Not selling any new services or not adding customers to the capacity limited site

Obviously, the rank ordering of the capacity-relief options listed previously can be changed, based on market as well as system conditions. Steps 8 and 9, although they are options, are not desired choices.

The first step in identifying the capacity cell sites required for a network involves the use of a spreadsheet to determine where the problems are anticipated.

The capacity-planning phase involves taking the marketing information and assigning a weighted value of the total system traffic to each sector. This process is done primarily by taking the usage information provided by marketing that has demographic information pertaining to the actual location(s) of the expected customers. Obviously, the current customers' locations are known, but marketing needs to provide any significant changes expected to traffic types, usage, and patterns expected. The objective of obtaining this information is to arrive at a Mbps/km² value, from which traffic projections can then be extrapolated.

One immediate problem that arises deals with the problem of having voice and data traffic using the same radio transport channel. Therefore, it is important to convert all traffic data into the same format, namely Mbps. The method for converting a voice channel to Mbps is straightforward, but appears illusive until shown how to do the conversion.

DS0 = 64kbps = 0.064Mbps

Erlang = 1.536 Mbps (T1) or 1.92 Mbps (E1)

This conversion, although straightforward, does not address the wireless or wireline transport efficiency. Care should also be taken when determining capacity requirements for a system because of the plethora of services that can be offered. Some quick examples involve IP services and whether you oversubscribe or provide no overbooking of the bandwidth. There are, of course, multiple variants to the IP example used, but the point about being cautious should be headed. Another simple example involves UBR, CBR, rt-VBR, and non-rt-VBR; and the many variants to the vast amount of services that utilize each of these classes of services.

Therefore, what needs to take place is a detailed marketing analysis (or as detailed as possible) with the output being the type of services you need to offer, their bandwidth or Mbps requirements, and different SLAs that could impact the capacity calculation. Although this may seem straightforward a few minutes into defining what you are going to sell, how it will be sold and the take rate for a given price becomes a daunting process. As has been the situation in LMDS system deployments and other wireless data systems, the barrier to entry has been significantly underestimated. In addition, the miscalculation of what the market is willing to bear and the ever-changing conditions have played havoc on many operators.

With that said, many times the RF engineer is often asked to provide cursory information regarding a market with regards to the capital (base stations, for example) that will be required to be deployed.

Example A quick system analysis that can be performed to determine how capacity is expected for the system to carry or need can be determined using the following:

R = Radio Spectrum Available = 50MHz (duplexed)

N = Number of channels = 4

M = Mbps/channel = 14 Mbps

S = Nominal number of sectors per hub site = 4

Z = Maximum radios per sector = 2

E = Efficiency factor (how much you want to load the channels) = 0.8 (80%)

C = Capacity required to be carried for the given market area = 1.5 Gbps during the defined busy hour

Therefore, the number of hubs or base stations required is determined using the following equation:

$$\text{Hub Sites} = C/(S \times Z \times M \times E)$$

$$1.5\text{Gbps}/(4 \times 2 \times 14\text{Mbps} \times 0.8)$$

$$= 16.74$$

$$= 17$$

Using the previous quick calculation for capacity, only a total of 17 base stations, hubs, are required to meet the traffic demand. The quick-capacity analysis, however, does not account for coverage issues and assumes that all the traffic is homogeneously distributed.

Coverage considerations can be swagged with the previous equation by altering the efficiency factor to say go to 40% as one quick suggestion.

The quick calculation for hub sites is used primarily for initial build evaluations when the issue is often raised about how many sites will we need to build.

Another analysis that is often asked for is to determine the impact to the required sites if a different modulation scheme is used for the same spectrum.

Example If the modulation scheme used were increased to, say, 64QAM (*Quadrature Amplitude Modulation*) from 4QAM what is the impact to the required sites for the system? Using the previous example numbers and assuming the same reuse factor, channel bandwidth and reuse plan a simple substitution can take place.

Previously M=14Mbps, now it can be assumed that M=25 Mbps.

$$\text{Hub Sites} = C/(S \times Z \times M \times E)$$

$$= 1.5\text{Gbps}/(4 \times 2 \times 25\text{Mbps} \times 0.8)$$

$$= 9.375$$

$$= 10$$

Therefore, increasing the modulation scheme resulted in a reduction of 58% of the sites required. However, with increased modulation comes reduced range and sensitivity, resulting in a smaller footprint for the site, reach. The result is that based on how disbursed the customers are, the analysis, although gross, can significantly mislead due to the range problem that is now compounding the issue.

But to deploy a system properly, a more in-depth analysis needs to be performed. The methodology for arriving at the amount of Mbps is shown in another chapter in sufficient detail. Because engineering is provided to the traffic calculations or forecast from marketing (at least that how it should work), the following process is based on the traffic data being provided to engineering for the proper dimensioning steps to take place.

If marketing is unable to determine the capacity calculations, the methodology and process used in Chapter 4 will resolve that gap, or at least minimize it.

Therefore, when performing the design effort, the first step in determining the amount of capacity that will be carried by any cell or sector in a PMP system is to identify where the bulk of the traffic for the system will originate. Because the traffic is stationary, the location of the subscribers eliminates many of the variables that are normally associated with a wireless system-capacity plan.

The marketing data provided should have a breakdown of usage, Mbps, for both voice and data for each geographic region. The reason for the differentiation is the packet- and circuit-switch approaches needed for designing the capacity requirements. For this part, the design can either assume that DBA will be available or will not be available.

If DBA is assumed to be available, the differentiation of voice and data is not that important because it will be factored into the Mbps usage during the BH. However, if DBA is not available, then voice usage will have to assume fully nailed-up connections and the exhaustion of the spectrum.

In addition, the potential to have T1/E1 traffic as an alternative backhaul platform will need to be factored into the capacity design. The QoS differences are not factored into a system design model, but need to be factored into the individual base station site's design and once the real take rates for each service in a market are better understood.

Utilizing a simple grid system, as depicted in Figure 5-10, for a market or portion of that market, each of the grids will have a particular amount of Mbps assigned to each site grid. Once the amount of traffic or load carrying is required for each of the grids, that traffic can then be assigned to a particular site, sector, or portion of a sector, depending on the granularity of the grid system utilized.

The grid system chosen could be based on *Uniform Transverse Magnetic* (UTM) or Lat/Long coordinates, Zip codes, or some other geocoded method. It is advisable that the grid system chosen by the engineering team be the same that the one marketing and sales is using to facilitate the planning process.

Effectively, you are putting a weighting factor for each sector that provides some level of coverage to a grid. The specific weighting is achieved through establishing the percentage of each grid that a sector of a site covers. For instance, if there are three sectors that cover a particular grid, one

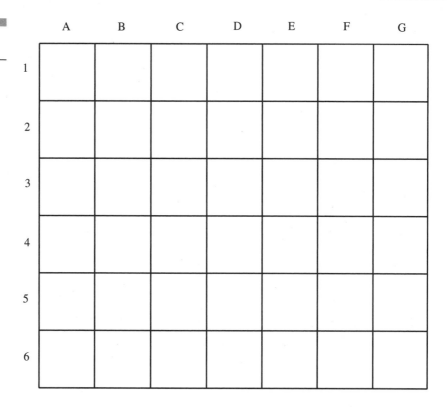

Figure 5-10
LMDS grid

sector may only provide 30%, another sector 10%, while the third provides 60%. Therefore, using the previous numbers, the traffic for each grid is then distributed to each of the sectors in proportion to their coverage of that grid.

The advantage of this type of approach is that because the sites are fixed, it is possible to perform multiple sensitivity studies on the impact, varying the type of services offered will have on the capacity-carrying ability of the system as well as to an LMDS site level.

This distribution is maintained for each grid unless a new cell or sector is added to the area, thereby requiring a readjustment of the traffic distribution for that time period. The addition of a new site or sector could be a result of the marketing rollout plan or a result of capacity relief requirements.

Now, to apply the before-mentioned information into practice.

From marketing, the following information is provided. More detail, of course, can be extracted, but that is dependent upon the services offered, the services take rates, and the technology platform that will be used. Table 5-6 shows different service classes that can possibly be used for the traffic calculation. For the example, any oversubscription has been factored into

Table 5-6

Traffic projection
template

Type (GRID A1)	Services	Year 1 Mbps	Year 2 Mbps	Year 3 Mbps
Fixed (CBR and PVC)	T1/E1			
	64K			
	128K			
	256K			
	512K			
	1M			
Variable CBR	(BRI+D)			
	PRI			
Non-rt-VBR	512k			
	1M			
UBR (IP)	256k			
	512K			
	1M			

the capacity calculation as UBR-type traffic. The obvious question that may arise is, why not define some VPN/WAN traffic types such as Frame Relay and ATM, etc.? The reason is that for the purpose of determining the radio link capacity, the specific type of service, although extremely interesting, is not really needed. For instance, if a PVC is set up, it is constantly in place from the radio system's aspect and it does not matter if it is transporting voice, IP, Frame, ATM, etc.

Clarifying this, the traffic project is to determine the amount of Mbps/km^2 for any geographic area, which then can be used to determine the amount of infrastructure required. However, you can easily expand or constrict the table to show the required information needed.

The determination about which services are offered through the marketing and sales effort should be driven by the marketing and sales department, not engineering, except when there is a technological issue. For example, leased line replacement is being sold but the infrastructure design is meant for IP traffic and is a layer 3 router is an example of when a direct conflict happens.

Continuing, I picked three end points to use; those will be different, depending on what the design requirement calls for.

The fixed (CBR and PVC) information can refer to lease line replacements, fractional leased lines, Frame Relay, ATM circuits, and voice circuits (POTS). Variable CBR is shown for ISDN services, whose activity can be determined via the D channel. Non-rt-VBR could be batch file transfers for a simple example. The UBR (IP) is more classical Internet IP and not VoIP, at least not yet. If VoIP is desired between branch offices, Voice over Frame may be a better solution at this point.

The information in the table needs to be established on the same grid area to facilitate the traffic spreading needed for a realistic traffic analysis. Therefore, the data needs to be defined on a grid level so we can arrive at a Mbps/km^2 value from which the amount of radio channels can be derived.

Figure 5-11 is an example of what the desired output of the traffic grid system can show. From the marketing demographics, the expected customers are located in the grids shown. The next step is to place the base stations over the grid system, which will minimize the initial capital required but also position the system for growth into the future.

Figure 5-11
Mbps/km^2

	A	B	C	D	E	F	G
1	5						
2			13			6	
3			3.6	31	8.2		
4			2	0.5			
5							
6						5	

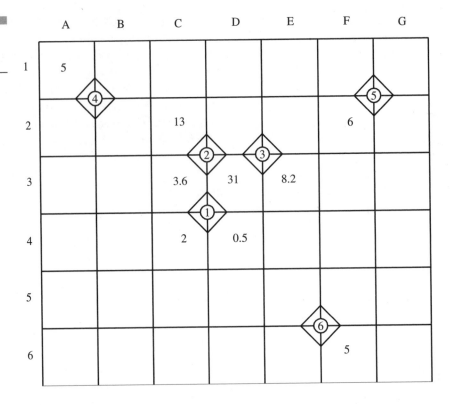

Figure 5-12
LMDS initial system
layout

Depending on the size of the grid, a base station could cover the entire grid or a portion of the grid. For the previous two-dimensional design, the sites will be placed on corners of the grids for simplicity.

The grid in Figure 5-12 shows that six sites are placed. Sites 1, 2, and 3 comprise the core of the service area initially targeted. Placement of three sites is not required from the grid aspect, but three were used to facilitate coverage. Site 4 may only have one sector active initially, as is the case with sites 5 and 6.

This effort can now be carried out for the remaining time periods for the forecast.

Now, what do you do when the system is already operational? Well, the primary process is effectively the same. Specifically, a revisit to the existing services offered, the capacity currently being carried by each of the grid areas, which can be provided by measured traffic on the system. Then, the marketing and sales forecast can then be added.

Table 5-7

Traffic projection

Type (GRID A1)	Services	Current Mbps	Year X New Mbps	Year X Total Mbps
Fixed (CBR and PVC)	T1/E1			
	64K			
	128K			
	256K			
	512K			
	1M			
Variable CBR	(BRI+D)			
	PRI			
Non-rt-VBR	512k			
	1M			
UBR (IP)	256k			
	512K			
	1M			
Total		5Mbps	7Mbps	12Mbps

The difference in methodology is reflected in Table 5-7. Total Mbps is arrived at using the following formula:

Total Mbps = Current Mbps + New Mbps

The method is to measure what the current level of usage is for the BH time period and then use marketing's forecast for the additional services (Mbps) that will be added for each grid. The result is obviously the present usage plus the new usage. It should be stressed that the delta service offering could be negative if a particular service currently offered is discontinued in a geographic area due to cost or other reasons.

Frequency Planning

Frequency planning, or frequency management, is an integral part of any wireless system. For all RF engineers, LMDS systems involve frequency planning and many of the processes are similar to mobility planning with the one key concept: the customers are stationary. As with any frequency plan, the objective is to maximize the utilization of spectrum. The maximization is achieved through reusing the spectrum, if required, so that more capacity, or rather more Mbps/km², can be carried for any geographic area.

How much frequency planning there is in a network is largely determined by the technology platform that is chosen by the operator for use. There are many variants to frequency planning, ranging from coordination of a single transmit channel to orchestrating the manipulation of hundreds of radio channels. All too often, it is the frequency planner who sets the direction for the performance of a communication system. The frequency planning process needs to be rigorously checked on a continuous basis to always refine the system. As a minimum for frequency planning, the designs, no matter how minor they seem, need to be passed through a design review process.

The use of different bandwidths and modulation schemes results in a plethora of C/I ratios that equate directly into D/R requirements. Added to this is the simple problem of limited spectrum—no matter how much spectrum an operator has, there is always a need for more spectrum for capacity, system flexibility, and performance while reducing the cost of delivering an Mbps to the end user.

The use of a grid is essential for initial planning, but when "sprinkled with reality," the notion is academic in nature. The reason it is academic is because the site acquisition process tends to drive the system configuration, not the other way around. It is dealing with the irregularities of the site coverage, traffic loading, and configurations that require continued maintenance of the network frequency plan.

But a grid is where the system design needs to begin from a layout aspect in order to facilitate reuse within the system. The LMDS grid shown in Figure 5-13 consists of a series of points that base stations can be placed on. The spacing is done here with an overlap between adjacent cells. The grid pattern shown starts off with two sites that are geographically adjacent, but do not need to be. The sites are placed to allow for the expansion of another base station in the future.

The radius for the base station is extracted from the efforts done when the link budget is calculated. And, of course, the distance between reusing sites is a function of the D/R ratio.

Figure 5-14 shows the relationship between distance between reusing sites and the radius of the site itself.

Figure 5-13
LMDS grid

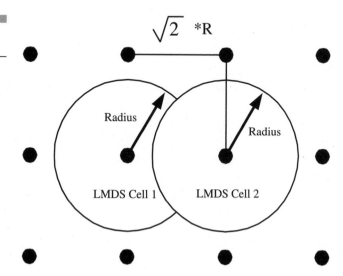

Figure 5-14
D/R ratio

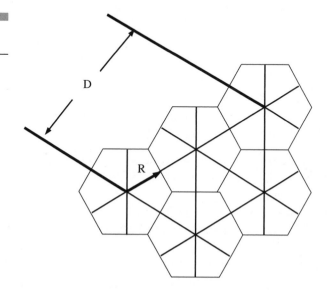

What are some of the options available to the design engineer? The choices are vast and depend on some obvious issues like the following:

- Spectrum available
- Channel plan used
- Modulation format
- C/I requirements

Under ideal conditions, there would be no frequency reuse because there would be an infinite pool of channels to draw upon. However, you are always placed with some simple restrictions, normally starting with available spectrum, which may or may not be channelized.

With any LMDS system, there can be two polarizations used in the system, *Vertical* and *Horizontal* (V and H). It is essential that the infrastructure utilized supports both polarizations, which are really needed for capacity-expansion purposes.

There have been multiple frequency plans presented for LMDS systems, but most foster the use of dual polarizations within the same sector of a base station, or when sector splitting utilizes another polarization for part of the expansion. It is imperative, however, that you utilize the same polarization for the same sector, even when splitting the sector for the following general reasons:

- To prevent reconfiguration of host terminals when a frequency change is needed
- Traffic tailoring is possible through moving host terminals, based on their SLAs, to selected channels within the same sector to improve the bandwidth utilization
- To facilitate the capacity expansion of the sector itself
- To reduce capital expenses and spare volume requirements

The following frequency reuse plans should address most, if not all, system requirements. The frequency plans do not, however, address a dual band system because each band should be designed separately with its own set of frequency reuse conditions.

The patterns presented represent the following general classifications of reuse. The reuse patterns shown can work with a host of channel bandwidths, in addition to technology platform types. Associated with most of the patterns is an example of how to handle capacity issues that will arise within the system due to success problems.

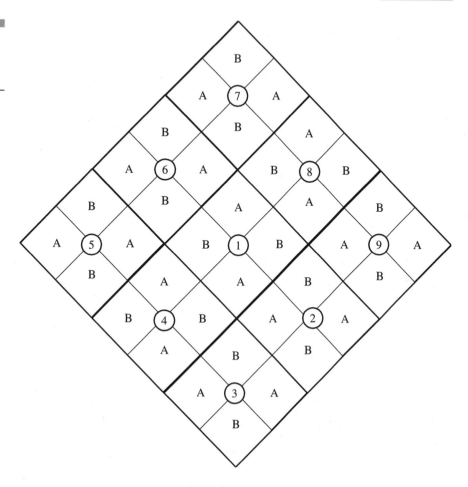

Figure 5-15
Two-channel—same
polarization
frequency plan

- Two-Channel—Same Polarization
- Two-Channel—Dual Polarization
- Three-Channel—Dual Polarization
- Four-Channel—Dual Polarization

Two-Channel—Same Polarization

Figure 5-15 is a frequency reuse pattern that should be used when presented with only two active channels that can be used for the system design effort. In addition for this plan the ability to use alternating polarizations is not feasible.

Figure 5-16
Two-channel—dual
polarization
frequency plan

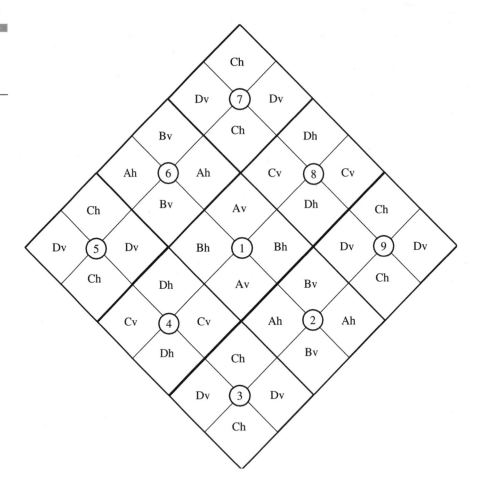

Two-Channel—Dual Polarization

Figure 5-16 is a frequency reuse pattern that should be used when presented with only two active channels that can be used for the system design effort where the use of alternating polarizations is possible.

Figure 5-17 shows how expansion is possible with the use of sector splitting when utilizing the frequency plan shown in Figure 5-16.

Figure 5-17
Two-channel—dual
polarization sector
split

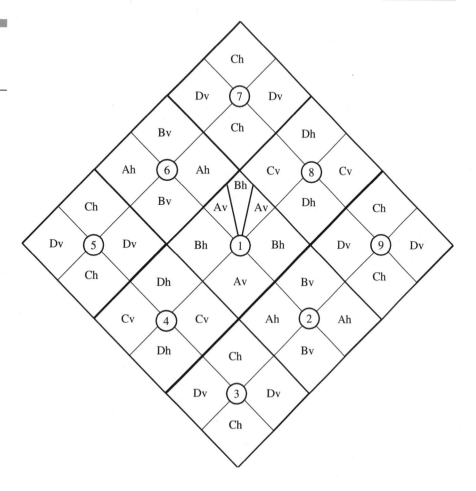

Three Channel—Dual Polarization

The frequency plan shown in Figure 5-18 can be used when only three channels are available for use by the system operator.

Four-Channel—Dual Polarization

Figure 5-19 shows a frequency reuse pattern that should be used when presented with four active channels that can be used for the system design effort. In addition to this plan, the ability to use alternating polarizations is available.

■ ■ ■ ■

Figure 5-18
Three-channel—dual
polarization
frequency plan

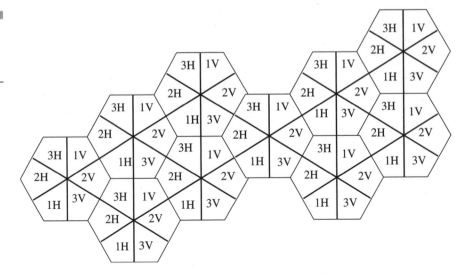

Figure 5-20 represents how capacity expansion can take place using channel additions with this reuse plan, through the addition of radio channels to any sector.

The next two figures show two distinct but related methods for expanding the traffic-carrying capacity for the site. The expansion plan can be done in stages, as follows:

1. Add radio channel

2. Increase modulation scheme for one radio

3. Split the sector

This is, of course, provided that there are not a few heavy users who could be migrated to a PtP connection instead.

For the expansion example shown in Figure 5-21, this process could occur in place of a sector split. If the underlay overlay scheme were used, then the outer ring would utilize the lower-modulation scheme while the inner ring would utilize the higher-capacity modulation method. Host terminals then could be moved between radios in order to increase the utilization and throughput for that sector provided the same polarization was used for all the host terminals in that sector.

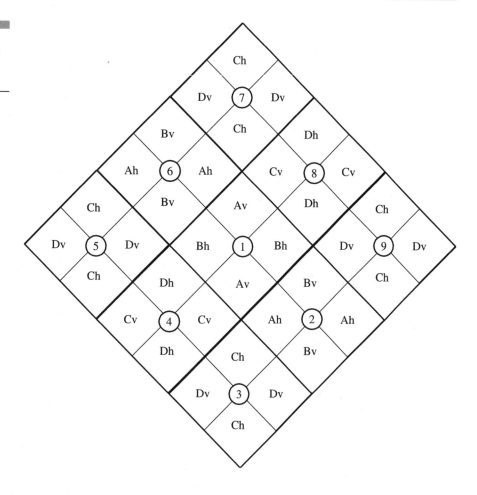

Figure 5-19
Four-channel—dual
polarization
frequency plan

The sector split shown in Figure 5-22 could be accomplished with higher modulation schemes and/or after the addition of a higher modulation method is employed. That is, provided that the infrastructure can support higher modulation methods and it is achieved via a software change.

Figure 5-20
Four-channel—dual
pole frequency reuse
channel expansion
plan

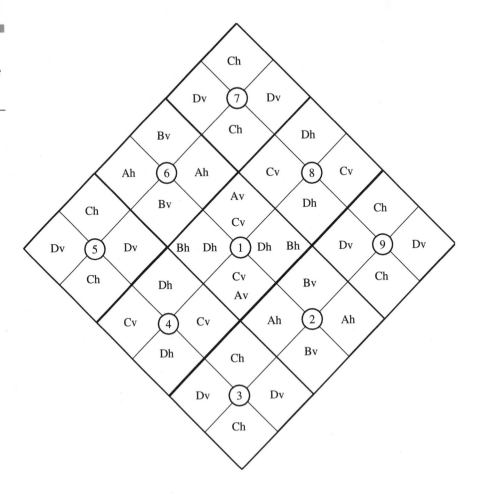

Figure 5-20
Four-channel—dual
pole frequency reuse
channel expansion
plan

Propagation

Every wireless system sometimes requires a propagation analysis to be performed. In wireless mobility, the use of propagation prediction tools is used extensively. LMDS systems are no exception to the rule of requiring a propagation analysis. The primary difference lies in the type of propagation analysis.

Typically, a wireless system will use a propagation-analysis tool to design the initial location of base stations and perform interference analysis to support its frequency reuse plans.

In LMDS, the propagation tool is used to determine where in a given region around a base station service can be delivered by the operator. The

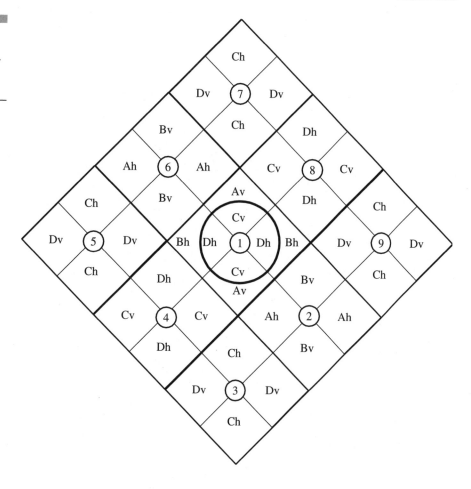

Figure 5-21

Four-channel—dual polarization capacity expansion with underlay overlay

result of this information is to augment the sales department by helping prioritize or identify regions within the system, one or many base stations, in which a host terminal can be located and properly connected to the system.

LMDS propagation analysis, however, is largely driven by LOS prediction, not multipath propagation. However, in the lower bands (below 3GHz and especially below 1GHz), the use of multipath can be taken advantage of.

Typically, for LMDS deployments, the use of LOS is the predominant method of calculation for coverage. The LOS propagation equation is the classical free space equation:

$$PL \text{ (free space)} = 32.4 + 20\log(R) + 20\log(f)$$

where R = km and f = MHz [a] smith

Figure 5-22
Four-channel—dual
pole sector split

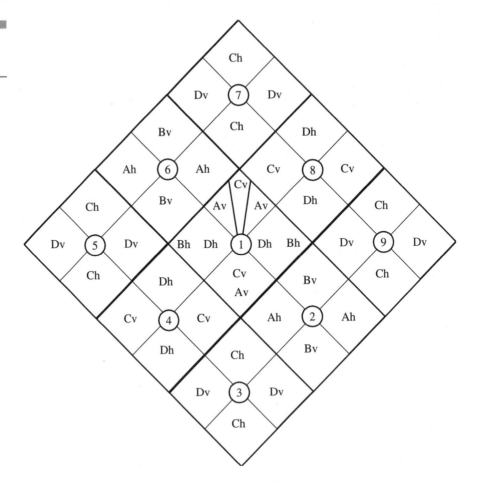

There are several key elements that need to be factored into the free space equation, which is where propagation tools come in handy. The first area is the inclusion of terrain and earth curvature. The better the terrain database, the better off the design will be. The other key criterion is the inclusion of clutter, or, more specifically, building height data.

Many debates and much money have been spent on utilizing a detailed building height data base in order to properly locate the base stations and determine where the most potential host terminals can be reached. In practice, however, the building height database is always less than stellar because the cost is so high for any city; the use of one- or five-meter resolution still results in concerns for obstructions. The natural choice is to utilize the one-meter database, but this data will not accurately reflect HVAC installations, window-washing equipment, other wireless installation, etc.

The point I am making is that even with the best resolution data base, a site survey still needs to take place. An LOS analysis, using a pair of binoculars, camera, handheld GPS, and a good map will probably result in better, cheaper, and more expedient information. Also, the location of the base station that is chosen may not be the ideal site, but it can be leased and can cover about 80% of what was desired from the initial design process.

Nevertheless, propagation analysis is still valid to perform for a network design in order to determine the rough area that the site can possibly cover. Utilizing building height information obtained from various site acquisition firms the suitability of a site can be determined.

An LOS propagation analysis for an unobstructed site looks like what is shown in Figure 5-23.

All too often, when management comes from a mobility or two-way environment, they expect the propagation analysis to reflect that shown in Figure 5-24.

In reality, the only reason why an LMDS site that requires LOS is not circular for propagation prediction is due to terrain obstructions or buildings (if the database was purchased and used). Therefore, the best thing that can be done for the design process that involves propagation predictions is to use the tool to determine approximately the sites reached without man-made obstructions, and then use this data to indicate on a map the areas supported with radio coverage for the sales teams.

The propagation model, however, has far more usefulness in the lower bands, in which multipath communication is possible. One such example is the UHF band in the United States. The UHF TV channels are now being made available for broadband services. For this area of analysis, an LOS approach is good, but the use of a propagation tool is more effective. Specifically, the Hata model for propagation is amply suited for use in this band.

The Hata model is the most prolific path loss model employed in wireless communication, particularly cellular. The Hata model is an empirical model derived from the technical report made by Okumura, so the results could be used in a computational model.

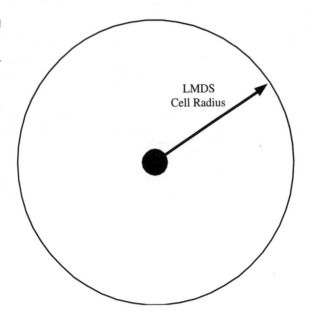

Figure 5-23
LMDS LOS radius

LMDS
Cell Radius

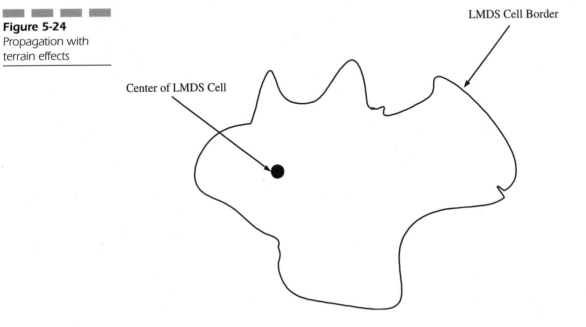

Figure 5-24
Propagation with
terrain effects

LMDS Cell Border

Center of LMDS Cell

RF Design Implementation Guidelines

RF Engineering for an LMDS site or system cannot take place without some form of RF design guidelines, whether formal or informal. With the level of complexity rising everyday in the wireless communication systems, the lack of a clear definitive set of design guidelines is fraught with potential disaster. Although this concept seems straightforward and simple, many wireless engineering departments, when pushed, have a difficult time defining what exactly their design guidelines are.

The actual format or method of the way the guidelines are conducted should be structured to facilitate ease, documentation, and minimization for formal meetings. For most of the design reviews, a formal overhead presentation is not required; instead a sit-down with the department manager is usually the level of review that is needed. The important point is that another qualified member of the engineering staff should review the material to prevent common or simple mistakes from taking place. Ensuring that a design review process is in place does not eliminate the chances of mistakes taking place. Design reviews ensure that when mistakes do take place, the how, why, and when issues needed to expedite the restoration process are already in place.

It is highly recommended that the department's RF designs guidelines be reasonably documented and updated on a predetermined basis—yearly at a minimum. The use of design guidelines will facilitate the design review process and establish a clear set of directions for the engineering department to follow. The RF design guidelines will also ensure that there is a consistent approach to designing and operating the capital infrastructure that has or will be put into place within the network.

The actual design guidelines that should be utilized by the RF engineers need to be well-documented and distributed. The design guidelines however, do not need to consist of voluminous amounts of data. They should consist of a few pages of information that can be used as a quick reference sheet by engineering. The design guideline sheet has to be based on the system design goals and objectives set forth in the RF system design.

The actual content of the design guidelines will vary from operator to operator. However, it is essential that a list of design guidelines be put together and distributed. The publication and distribution of RF design guidelines will ensure that there is a minimal number of RF design specifications in the network.

The proposed RF design guideline is shown in Table 5-8. The design guideline shown is for an LMDS system and can easily be expanded upon.

Table 5-8

LMDS RF design
guidelines

System Name
Date:

Frequency Range of Operation:

Spectrum Allocated:

Band Plan (if applicable):

Channel Bandwidth: 12.5MHz and 25Mhz

System Type: TDMA, FDD, TDD

Tx/Rx Spacing: MHz

	Modulation	Bandwidth	Rx Sensitivity	Mbps
Radio Type A	4QAM	12.5MHz	−82dBm	14Mbps
Radio Type B	64AM	12.5MHz	−72dBm	28Mbps
Radio Type C	4AM	25 MHz	−78dBm	25Mbps

Radio Type Deployed:

C/I or Eb/No: 15 dB

Frequency Reuse   N=0.5

Nominal Number Channels per Sector:

Maximum Number Channels per Sector:

Polarizations: Vertical, Horizontal, Circular

Orientation of Sector 1: degrees TN

Base Station Antenna Gain: dBi

Host Terminal Antenna Gain: dBi

Base Station EIRP:

Host Terminal EIRP:

Host Terminal Rx Sensitivity:

Range of Site (km or miles):

Rain Zone:

Fade Margin (rain attenuation):

Link Reliability:

Cell Overlap: dB or %

Performance Criteria

Link Reliabilty: 99.999%

Bandwidth Utilization: 70%

CBR/UBR ratio:

The guideline can easily be crafted to reflect the particular design guidelines utilized for the market where it will be applied.

Base Station Design

Although this is not necessarily the first step in any design process, it is one of the most important for the RF Engineering department. The reason the site design is critical is that it is where the bulk of the capital is spent. The LMDS cell site design guidelines listed as follows can be utilized directly or modified to meet your own particular requirements.

The use of a defined set of criteria will help facilitate the site build program by improving interdepartmental coordination and provide the proper documentation for any new engineer to review and understand the entire process with ease. Often, when a new engineer comes onto a project, all the previous work done by the last engineer is reinvented primarily because of a lack of documentation and/or design guidelines from which to operate.

The cell site design process takes on many facets and each company's internal processes are different. However, no matter what the internal process is, the following items are needed as a minimum:

- Search Area
- Site Acceptance
- Site Rejection
- Regulatory Guidelines
- Aeronautical Obstruction Guidelines
- Planning and Zoning board
- EMF Compliance

Search Area

Defining a search area and the information content provided is a critical first step in the realization of LMDS site design. The search area request is a key source document that is used by the real estate acquisition department of the company or its outsourced vendor. The selection and form of the material presented should not be taken lightly because the RF engineers often rely heavily upon the real estate group to find a suitable location for the communication facility to exist. If the search area definition is not done properly in the initial phase, it should not be a surprise when the selection of candidate properties is poor.

The search area's issued have to follow the design objectives for the area of the system, following the RF system design objectives. The search area should be put together by the RF engineer responsible for the site's design. The final paper needs to be reviewed and signed by the appropriate reviewing process, usually the department manager, to ensure that there are checks and balances in the process. The specifications for the search area document have to not only meet the RF Engineering department's requirements, but also those of the real estate and construction groups and the network engineers. Therefore, the proposed form that follows needs to be approved by the various groups, but issued by the RF Engineering department.

It is imperative that the search area request that is issued undergo a design review prior to its issuance.

The proposed format that should be followed for the search area request is shown in Table 5-9.

Referring to the search area request form, the following comments need to be made. The map included with the search area request needs to include as much information as is practical for the real estate acquisition group to help locate the proper sites. The map used in this form will minimize the amount of dud sites that are presented to RF Engineering for consideration in the system design. The search area request map needs to include area-specific information.

This area-specific information varies from location to location. The variations that you can use for the map format are largely dependent upon your design criteria for the site. If the search area ring is very defined, as is the case with mature systems, then it is imperative that the adjacent existing sites and search areas be identified on the map itself. The rationale behind including adjacent sites and the current search areas will better define options available to the real estate acquisition group.

A propagation plot and/or desired coverage area for the search area request form needs to be also generated and included with the search area folder maintained by RF Engineering. The objective behind having a defined coverage area objective generated is to help define the search area and coverage rings put forth in the map provided. The propagation plots or coverage objectives will be used as part of the site-acceptance procedure listed later. The coverage area definition is one of the steps taken to ensure that the proposed site meets its desired objective.

Search Area Request Form

Search Area Code:_____ Capital Funding Code:_____

Issuance Date:

Target On-Air Date:

Search Area Map:

Number Sectors:

AGL:

ASML:

Size of Equipment Room: (ft² or m²)

Number Antennas:

Maximum Cable Length (ft, m):

Comments:

Search Area Request

Document Number _____ Date: _____

Design Engineer:_____

Reviewed by:_____

Revision:_____

On the search area request form, the search area code should be identified, along with its capital funding number. The capital funding number and the search area code should be the same.

The on-air or system-ready target date is meant to identify when the site needs to be placed into commercial service. The on-air date should match the dates put forth in the RF system design plan. The on-air date's purpose is to help prioritize the internal resources of the company by helping define the importance of the site.

The total number of antennas and their type are also important to define in advance for the site search. If a site requires three antennas versus sixteen antennas, the discussions that the real estate acquisition group takes with the potential landlord will be different.

The rationale behind listing the *Above Ground Level* (AGL) and *Above Mean Sea Level* (AMSL) values is to define the site's location options for the real estate acquisition group. The AGL will help structure the search for suitable properties that fit the design parameters. Sometimes, sites are available that are 10 meters tall and have a willing landlord and permit accessibility. However, the design specification calls for a 40-meter antenna height, which disqualifies the site for consideration and resource expenditure. If a site needs to be 30 meters tall (about 100 feet), its AMSL is important to define. It is important to define because it might be possible to have the antennas at the 30-meter height, but the location of the property is in a gully, requiring a 50-meter tall antenna installation.

The comment section of the search area request form is meant to provide an area for the designer to specify any particulars desired for the site.

On the search area request form itself is a section for documentation control. This section will track who issued the search area request, what revision it is, and the dates associated with it. Also on the document control portion of the form is a section for the design reviewer to sign off on the request.

It cannot be overemphasized that the information on the search area request form will largely define the success or failure of the property search. All search area requests need to undergo a design review.

Site Acceptance (SA)

It is strongly recommended that the RF engineer responsible for the final site design visit the site prior to site acceptance. This site visit will facilitate several items. First, the engineer will now have a better idea of the potential usefulness of the site and its ability to be built, and will provide more accurate instructions to the implementation team.

Once a site has been evaluated for its potential use in the network, it is either determined to be acceptable or not acceptable. For this section, the assumption will be that the site is acceptable for use by the RF Engineering department as a communication facility. It is imperative that the desires of the RF Engineering department be properly communicated to all the departments within the company in a timely fashion. The method of communication can be done verbally at first, based on time constraints, but a level of documentation must follow that will ensure that the design objectives are properly communicated.

The form listed in Table 5-10 is meant as a general guide to be used, but it might need to be modified, based on your particular requirements. Before the SA is released, it is imperative that it go through the design review process to ensure that nothing is overlooked. The SA will be used to communicate the RF Engineering's intention for the site and will be a key source document used by real estate, construction, network engineering, operations and the various sub-groups within engineering itself.

The SA will also need to be sourced with a document control number to ensure that changes in personnel during the project's life are as transparent as possible.

The proposed *site acceptance form* (SA) is included as follows for reference.

The SA form can be easily expanded upon to ensure that all the relevant information required within the organization is provided. Whatever the form or method ultimately utilized, it is important to include the information listed previously as the minimum requirements.

Most of the information included in the SAF is self-explanatory. It is imperative that, as with all the other steps in the design process, a design review and signoff take place, establishing a formal paper flow. The SA needs to include with it a copy of the predicted propagation used to generate the search area request. The SA also needs to include with it a copy of the expected coverage area versus what the design objective utilized to approve the site for RF Engineering.

The SA also needs to include a copy of the proposed antenna-installation configuration. The proposed antenna configuration is used by equipment, engineering, and construction to evaluate the feasibility of the proposed installation. The antenna configuration drawing is also used as part of the lease exhibit information.

A copy of the SA and the supporting documents need to be stored in a central filing location that is secure, preferably on a server. The use of a central filing location will enable all the information pertaining to this location and search to be stored in one location and not distributed between many people's cubes.

Site Rejection

In the unfortunate event that a potential LMDS site is determined not to be suitable for potential use in the network, a site rejection form needs to be filled out as shown in Table 5-11. The issuance of a site rejection form may seem trivial until there is a change of personnel and the site is tested again at a later date. The site rejection form serves several purposes. The first purpose is that it formally lets the real estate acquisition team know that the site is not acceptable for engineering to use and they need to pursue an alternative location. The second purpose is that this process identifies why the site did not qualify as a potential communication site.

Table 5-10

Site acceptance form

Site Acceptance Form

Search Area Code: _____ Capital Authorization Code: _____

SAF Document Number:

Site Address:

Lat:

AGL

Long   ASML

Ptp Backhaul (Y/N):

Regulatory Issues:

PtP License Needed (Y/N):

PtP Frequency Secured (Y/N):

FAA Approval Attached (Y/N):

FAA Marking and Lighting Required (Y/N):

Site-Specific Information:

   Antenna Configuration Attached:

   Radio Equipment Location Defined:

   Network Equipment Location Defined:

   Equipment Room and Location sketch attached:

Radio Equipment:

Network Equipment:

   Antenna Structure (roof, tower, monopole, water tank):

   Equipment Room: (Prefab, Interior Fitup [TI], Exterior):

   Approx Cable Length (ft, m):

Type and Quantity of Antenna (include PtP)					
Sector	Type	Number	Polarization	Orientation	EIRP

Existing Transmitters On Structure (Y/N):

If Yes, state Freq, EIRP, call sign, and physical location for each antenna and service:

Qualification Information:

Coverage Objective Obtained (Y/N):

Percentage of Area Site Will Cover Versus Design Objective:

IMD Study Complete:

EMF Study Complete:

Site Particular Comments:

SAF:

Document Number Date:

Design Engineer:

Reviewed by:

Table 5-11

Site rejection form

LMDS Site Rejection Form
RF Engineering

Search Area code:

The (<u>name of test location</u>) was visited on (<u>date of test</u>) and did not meet the design criteria for the search area defined.

The location did not meet the design criteria for the following reasons (<u>state reasons</u>).

RF Engineer: _____

Engineering Manager: _____

It is recommended that the site-rejection process include a design review with a sign-off by the manager. This is to ensure that the reasons for rejecting the site are truly valid and the issues are properly communicated. The form proposed in the SR needs to be distributed to the same parties that the SA would be sent to.

The fact that a site does not meet the design criteria specified at this time in the network design does not mean it will always be unsuitable. Therefore, it is imperative that the site survey information collected for this site be stored in the search area's master file. The storing of the site survey information in the central file will assist efforts in later design issues that could involve capacity or relocation of existing sites to reduce lease costs.

Site Activation

The activation of a cell site into the network is exciting. It is at this point that the determination is made for how effective the design of the cell site is. There are numerous steps, after the site acceptance process previously listed, that need to take place. The degree of involvement with each of these steps is largely dependent upon the company resources available and the interaction required between the engineering and construction departments.

At a minimum, these two groups should perform site visits together. These site visits involve the group responsible for the cell site's architectural drawings and overall design of the site's structure.

Regardless of the interaction between the groups when it comes to "show time," it is imperative to have a plan of action to implement.

Regulatory Guidelines (FCC)

The FCC guidelines are unique for each type of radio service deployed. The guidelines set forth should be known by the design engineer and the technical management of the company. All the FCC guidelines affecting the radio community are contained in various parts of the Code of Federal Regulations, Title 47, CFR 47. Which section of CFR 47 applies to the particular system at hand depends on the license that the operator is utilizing to provide service. Each different part of CFR 47 dictates what can and cannot be used for the particular service. However, with the continuous flow of regulator changes that take place, it is strongly advised that you obtain the most recent CFR 47 and status the FCC's web page for dockets.

Planning and Zoning Board

Depending on the land use acquisition process needed, preparing for a zoning or planning board should be part of the design review process. Not only is the presentation important, it might be possible to eliminate this process entirely with a modification to the original site design. There have been many actual cases where a modification to the site design would have eliminated the need to request a variance from the town and thus prevented massive delays in the site build program. Although this step seems obvious, the lack of checking local ordinances and incorporating them into the design process is rare and often a "forgotten child" in the design process.

It is recommended that the local ordinances for the site be included with the site source documents.

When it comes time to present the case of why the site is needed to the local planning and zoning boards, a well-rehearsed presentation is needed. It is recommended that the program be rehearsed prior to the meeting night to ensure that everyone knows who will say what and when.

Engineering's role in the process tends to focus on why the site is needed, and health and safety issues associated with *electromagnetic fields* (EMF). The items that should be presented or prepared for should include the following as a minimum:

- Description of why the site is needed
- Discussion of how the site will improve the network

- Drawing of what the site will look like
- Views from local residents
- EMF compliance chart
- EMF information sheets/handouts for the audience (if applicable)

Before the meeting, it is essential that the local concerns be identified in advance so they can be specifically addressed before or at the meeting. It is also recommended that the presentation be focused at the public and the board members, not just the board members.

As in every case, it is imperative that all the issues needed to launch a successful appeal for a negative ruling by the council are covered. The overall preparation for the meeting is essential because the comments made by the company employees or consultants are a matter of public record and will be used solely for the appeal process.

EMF Compliance

EMF compliance needs to be factored into the design process and continued operation of the communication facility. The use of an EMF budget is strongly recommended to ensure that you are in compliance. The EMF budget is essential to ensure personnel safety and government compliance. A simple source for the EMF compliance issue should be the company's EMF policy.

The establishment of an EMF power budget should be incorporated into the master source documents for the site and stored on the site itself, identifying the transmitters used, the power, who calculated the numbers, and when it was last done. As a regular part of the preventive maintenance process, the site should be checked for compliance and changes to the fundamental budget calculation.

The method for calculating the compliance issue is included in the IEEE C95.1-1991 specification with measurement techniques included in C95.3.

The EMF power budget should be signed off by the manager for the department and shared with the operations department of the company. An EMF budget needs to be completed for every cell site in operation and also for those proposed.

Site Activation

The philosophy of site turn-on for LMDS systems requires no real integrated activation within the RF environment, except for sector and cell splitting. The LMDS site from an RF aspect is virtually isolated, so the site can be activated when it is most convenient for operations and engineering.

In addition, because since the LMDS site does not interact with the adjacent sites, such as in the case of mobility, the activation of the LMDS site can take place during normal business hours (unlike other wireless systems where activation is done during the maintenance window).

However, it is essential that for every cell site brought into a network that a plan is generated for its introduction, and that the plan is then carried out.

The design reviews necessary for a new site activation into the network need to be conducted by several parties. There are several levels of design reviews for this process. The first level of design reviews involve the RF engineer and the network engineer discussing the activation plans and reviewing the plan of action put forth. The second level of design reviews involves having the manager of the RF engineering group sign off on the implementation design with full concurrence with the performance manager. The third level of design review involves reviewing the plan with the director of engineering and operations personnel to ensure that all the pieces are in place and that something has not been left out, such as who will do the actual work.

After the design reviews are completed, the MOP for the activations is released. During the design phases, the MOP should have been crafted and all of the involved parties informed of their roles. A sample MOP is listed as follows for comparison (obviously, the exact MOP for the situation is different and needs to be individually crafted).

Method of Procedure for New LMDS Site Activation

Rev -X Date

Pre-activation Process

Date

X-X-XX	New cell sites to be activated defined
X-X-XX	Project leader(s) defined and timetables specified, as well as the scope of work associated with the project

X-X-XX	Phase 1 design review (frequency planning only and RF Engineer for site)
X-X-XX	Phase 2 design review (all engineering)
X-X-XX	Phase 3 design review (operations and engineering)
X-X-XX	Phase 4 design review (adjacent markets, if applicable)
X-X-XX	Frequency assignment and site-specific parameter sheets given to operations
X-X-XX	New cell site integration procedure meeting
X-X-XX	Performance evaluation test completed
X-X-XX	Executive decision to proceed with new cell site integration
X-X-XX	Adjacent markets contacted and informed of decision
X-X-XX	Secure post cell site activation war room area
X-X-XX	MIS support group confirms readiness for post processing efforts
X-X-XX	Customer care and sales notified of impending actions

New Cell Site Activation Process—(begins X-X-XX at Time: XXXX)

Operations informs key personnel of new cell site activation results

Operations personnel conduct brief post turn-on tests to ensure that there are no post-commissioning problems

Operations manager notifies key personnel of testing results

Post Turn-On Process—(begins X-X-XX at Time:XXXX)

Voice mail message left from engineering, indicating status of new cell sites (time)

Database check takes place

Statistics analysis takes place

Voice mail message left from RF Engineering, indicating status of post turn-on effort (time)

Commit decision made with Directors for new cell site (time)

Post turn-on closure report produced

New site files updated and all relevant information about the site is transferred to operations

It is essential to always include a back-out procedure for cell site activations, in case of a major disaster. The escalation procedure should be defined in the MOP, and the decision to go or not to go needs to be at the director level, usually the engineering director or the operations director.

After the MOP is released and the design reviews are completed, it is essential that the potential new cell is visited by the engineers at various stages of the construction period. Prior to activation, it is essential that a *Pre-Turn-On* (PTO) take place. The PTO is meant to ensure that the site is configured and installed properly, so that when the site is activated into the network, the basic integrity of the site is known. The PTO procedure that should be followed is listed in Chapter 8, "Implementation."

Internal coordination involving a new site or sites being introduced to the network is essential. The MOP listed previously focuses on voice mail notifications to many groups inside and outside the company. However, it is essential that the activation of new cells and major system activities is announced to other departments in the company to inform them of the positive efforts being put forth by engineering and operations.

The primary groups to ensure that some level of notification takes place are as follows:

- Sales
- Marketing
- Customer Service
- Operations, Real Estate and Engineering
- Corporate Communications
- Legal and Regulatory

Primarily, the entire company needs to be notified of the positive events that take place. One of the most effective methods is through the company's internal voice mail system associated with their extensions. However, not everyone will have an individual extension that necessitates the availability of a voice mail account.

To ensure that all the people are notified of the new site's activation into the network, a series of communications can be accomplished. One method is to issue an email message to all the employees, notifying them of the new sites and any particulars about the intended improvements, if any, to the network. Another method is to slay the trees by issuing a memo to everyone in the company declaring the activation of the sites and the improvements that have arrived.

External coordination for new sites is as essential as internal coordination. Specifically, the neighboring systems need to know when you are bringing new sites into the network when it affects frequency coordination. The reason behind this effort is that your actions may have an unintended consequence on them, either positive or negative, which they need to know. In the same light, by providing your neighboring systems with new site activation information, the same level of communication can be reciprocal.

After the sites are activated into the network, it is essential that post turn-on testing begins immediately. It is important that the efforts put forth in this stage of the site-activation process receives as much attention as the design phases did.

The key parameters or factors that need to be checked as part of the post turn-on activities are as follows:

Site Configuration checks

Metrics Analysis

Network Connectivity

New Site Performance Report

1. Site Configurations from the switches point of view

   The objective here is to check all the site parameters for the site, compared to those intended for the initial design. What you look for here is a possible entry mistake or even a design mistake made during the design process. Usually, a fat finger mistake is found in this process or an entry is left out.

2. Metrics Analysis

   The objective of this part of the post-turn-on activities is meant to help identify and isolate problem-resolution problems reported in the network by the system statistics. This process requires continued attention to detail and an overall view of the network at the same time.

   The particular set of metrics used is directly dependent upon the infrastructure vendor used and the services rendered.

3. Network Connectivity

   The objective with this is to verify that the path of a bit of information, whether it is voice or data, from the time it enters into the LMDS system to when it leaves, is configured properly and undergoes the required treatments envisioned for the services offered.

4. New Site Performance Report

   The last stage in the new site activation process is the issuance of the new site performance report. The performance report will have in it all the key design documents associated with the new site. The key design documents associated with this new site should be stored in a central location instead of a collection of people's cubes.

The information contained in the report is critical for the next stage of the site's life. The next stage of the site's life involves ongoing performance and maintenance issues.

To ensure that poor designs do not continue in the network, it is essential that the new site meet or exceed the performance goals set forth for the network. If the site does not meet the requirements set forth, then it should remain in the design phase and not the ongoing system operation phase. The concept of not letting the design group pass system problems over to another group is essential if your goal is to improve the network.

The New Site Performance Report needs to have the following items included in it as the minimum set of criteria:

- Search area request form
- Site acceptance report
- New cell site integration MOP
- Cell site configuration drawing
- Frequency plan for site
- Other site parameters
- FCC site information
- FAA clearance analysis
- EMF power budget
- Copy of lease
- Copy of any special planning or zoning board requirements for the site

The new cell site performance report is an essential step in the continued process for system improvements. Only once a site is performing at its predetermined performance criteria should the site transition from the design phase to the maintenance phase.

Network Design Guidelines

Introduction

The LMDS network system design is unique for wireless systems because it may, depending on the services offered and the technology platform chosen, have multiple protocols to deal with. The use of multiple protocols for use within a telephony and wireless systems is not unique. However, the fundamental choice of which platform will be the predominant force in the network design can lead to many perplexing situations.

The complications can be further magnified, depending on the amount of on-net and off-net traffic the LMDS system will need to handle and the types of protocols they comprise. The back haul from the base station to switching/packet concentration node adds another wrinkle in this effort, based on traffic volume and interconnect facilities that are available, either in the time frame desired, the cost of using those facilities, or both.

The telecommunication industry is moving toward convergence of the plethora of service protocols. The choice of which protocol will be used (that is, migrated to) has yet to be determined. Legacy services will also need to be supported until convergence really can take place. The convergence referred to is ATM or IP as the leading platform. Both have their advantages and are used for different applications.

So, the perplexing question is this: Which platform will you use that will be future proof and not require additional capital investments in the future due to technology obsolescence? The answer to that question depends on which part of the network you are referring to. If the part of the network is the edge, then the convergence is toward IP. However, for the core of the network, the convergence is toward ATM. As always, the solution is not based on a single killer protocol but on the proper application of each toward obtaining the desired solution.

Because there are multiple platform decisions to make, this chapter will attempt to cover the issues the network designer will need to address.

New and Existing System Considerations

The considerations for a new or an existing system have similarities, but also have distinct differences. For the new system, the marketing and design teams are presented with the problem of guessing where the traffic will come from, the protocol best meant to service the traffic, and the volume referenced in Mbps. However, an existing system already has many of those questions answered and it is the relative match of what is actually happening to what the plans were that needs potential modification. The only comfort the design team can have is that whatever they design the system for it will be wrong (it cannot be perfect).

Some of the considerations that need to be reviewed for a new or existing network include the decisions to lease, build, or even buy one or more access providers. The access providers could be an existing CLEC or fiber franchise or another LMDS operator.

The decision to lease or buy the concentration node will also need to be weighed. Interconnect node locations are an essential consideration for any LMDS operator. One obvious decision is to colocate with the interconnect provider. However, which interconnect provider do you colocate with and what happens if they are merged with one of your competitors? Also, if you are colocated with one operator, the process to switch providers becomes more tricky, but not impossible.

A decision will need to be made regarding the packet- and circuit-switching network. In particular, the decision about whether to lease capacity from another provider, thereby expediting the time to market and reducing operating expenses-but all at the cost of control.

The most important decision that drives all the others relates to which services you intend to offer, either at system launch or in one or two years. The decisions made will dictate the network configuration, which, if not chosen well, will result in excess capital expenditures in the future to compensate for an incorrect decision.

Services Offered

The services offered by the LMDS operator need to be realized. One of the areas in which realization takes place is the network design. The network engineering design is interested in the type of services offered, where they originate from, where they terminate, if any treatment is needed, the provisioning and monitoring methods, and the CIR with an overbooking factor that will be promoted with each service.

Just what services can be offered depends largely upon the radio infrastructure and not the network engineering design. What I mean is that the network engineering design, being fixed, can support and deliver any protocol and service offering requested, given the time and resources to accomplish the task. However, if the interface to the customer via the host terminal is 10BaseT ports providing only UBR traffic, offering rt-VBR for video conferencing is not viable because of the RF environment, not the fixed network.

Supporting the services and what that means is again a vast area that has multiple meanings. For instance, the service offering could be for wholesale service providing only (that is, the pipe provider). Another example is a retail service provider, where the service is provided by the company directly to the customer. But what exactly is service? Is service the delivery of the bandwidth only or do you provide adjunct services to support the primary bandwidth provision (that is, sell CPE, cable the customer, configure their routers, provide ASPs, and so on)?

Often, the services offered will change with time due to the varying market conditions brought on by competition (the LMDS operator is one). Therefore, the platforms used by the network engineering design need to be flexible enough in design to account for the vast array of unknown changes. Although is seems a daunting task, in reality it is really a scaling issue.

The three major platforms that need to be supported are

- ATM
- IP
- TDM

With each of these platforms, the types of services that are to be offered will be proposed by the marketing department, in conjunction with assistance from the technical community. Associated with each service offered, the following primary information must be known for each:

Committed Information Rate (CIR)

Peak Information Rate (PIR)

QoS

Overbooking Factor

Fixed or Variable Billing

This information is really derivatives of the SLA that is offered, with the noted exception of the overbooking factor.

The direction that the industry is migrating to is IP at the end terminal location, with ATM as the backbone. The use of TDM circuits has to be considered in the interconnection method when connecting to another service provider for further delivery options. However, IP is the prevalent platform that will govern LMDS deployments.

TDM/IP/ATM Considerations

Well, how do you decide which platform to utilize: TDM, IP, or ATM? Associated with this, what is the dimension or proportion that each will be within the network? The answer is not simple because experience dictates that no one platform is a solution for all situations and requirements. Although that may be true, but it does not solve the issue.

There are several things to consider when selecting the platform used for the network layout:

- What services do you need to support, keeping in mind that the key to LMDS success is targeting a niche market and not being an I-can-do-everything provider?
- What is the protocol the host terminal will interface with the customer and its bandwidth?
- What is the protocol the base station will interface with for concentration or treatment at another location?
- What is the interface expected between the concentration node and the PTT/CLEC and what services can they provide or do you intend on utilizing?

Table 6-1 indicates the platforms types (that is, the protocols) that are best used for different types of services offered.

In the table, NA refers to the fact that the protocol does not have a standard method of supporting that particular service.

Table 6-1

Protocol efficiency comparison [2]

Protocol	TDM Circuit	Voice	Packet Data	Video
ATM	1st	1st	3rd	1st
IP	NA	3rd	2nd	2nd
Frame Relay	NA	2nd	1st	NA

A fundamental difference between ATM and IP and Frame Relay is that ATM uses fixed length cells, whereas IP and Frame Relay use variable length packets.

Of course, TDM can handle ATM, IP, and Frame Relay services (but not cost-effectively), in addition to realizing that unless you want to become the next PTT, which is your biggest competitor, building a TDM system may not be a desired solution.

With that said, however, voice services are still a large percentage of the telecommunication market and if your service offering reflects any use of voice services, there exists the potential that you will utilize a class 5 circuit switch for some of your offerings. The utilization of a class 5 switch can be achieved in many methods by either constructing the facility yourself or by leasing excess capacity from another service provider. The first has the desired control function, but at-large capital expenditure and associated operational issues. The former allows for quicker implementation for service offerings, but control is not straightforward as are feature transparency; and the recurring expense, due to leasing, is another issue that needs to be factored into the capital/expense equation.

TDM Switching

The role of the TDM switch in the network has grown, along with the importance and complexity of the network itself. The exact role that TDM switches will have in the future is unclear, but they play an important role in providing the last mile, kilometer access. Several major LMDS service providers offer TDM switching functionality, whereas others offer only IP. The industry migration from TDM to IP/ATM platforms should be considered heavily when moving forward with the TDM switching decision.

No matter what is said, however, TDM switching will continue to play a major role in telecommunications. The role TDM switches began with first was the manual exchange (also referred to as just a switch), followed by the rotary exchange, the crossbar exchange, and eventually the development of the modern electronic *stored program control* (SPC) exchange. Even among the newer switches, there are various designs and functional applications, depending upon who the switch manufacturer is and what specific role the switch serves in the network: central office, tandem, *Private Branch Exchange* (PBX), etc.

Regardless of its type, all these switches have the same basic function: to route call traffic and concentrate subscriber line traffic. Some of the more common switch concepts and designs are briefly described as follows, along with a few example network applications.

Switching Functions

There are numerous functions of the switch within the network; they can be categorized into three basic groups: elementary functions, advanced functions, and intermediate functions.

Elementary switch functions include the process of connecting individual input and output line circuits (trunks) within the switch itself and the ability to control the distribution of communication traffic across clustered groups of line circuits (trunk groups). The ability to interconnect individual line circuits allows the transfer of voice or data signals between end subscriber units or between network nodes to take place in a controlled, selective manner. The switches' ability to direct or route traffic between individual line circuits, based upon larger defined groups of line circuits, allows for more efficient and reliable control of large volumes of system traffic.

The more advanced functions are digit analysis, generation of call records, route selection, and fault detection. Digit analysis is the process of receiving the digits dialed by the customer, analyzing them, and determining what action the switch should perform based upon this information. The actions include attempting to place a call to another party, connecting them to an operator for calling assistance, providing a recorded announcement stating that the digits dialed were in error, etc. The process of generating a record or multiple records for any calling activity taking place within the switch is crucial to creating the corresponding billing records for these calls. In the end process, this results in the final bill being completed for the customer. Therefore, it is important that the switch produces an accurate account (record) of all call-processing activity it performs.

The route selection function directs all system traffic within the switch to a specific set of facilities (transmission circuit, service circuit, etc.), based upon routing tables developed and maintained by the equipment vendor and the system operator. Finally, the detection of errors or problems occurring within the switches' own hardware and software, plus the identification of failures with any of the interconnected facilities, is a required function of the switch to ensure the operating quality of the network.

An example of an intermediate switch function is monitoring subscriber lines. This function involves the completion of regularly scheduled checks of all line circuits interconnected to the switch for proper operation. Is the circuit still functional and able to carry system traffic? If a circuit is not performing properly, it is taken out of service and the fault detection function is notified to alert the operations staff. See Table 6-2.

Circuit Switches

Telephone networks utilize circuit switches for processing and routing subscriber calls. Circuit switching can be the space-division type, the time-division type, or a combination of these two designs. These designs are utilized in the matrix of the circuit switch. (The matrix is where the actual switching of line circuits or trunks takes place.)

Space-Division Switching

In space-division switches, the message paths are separated by space within the matrix (thus, the derivation of the name). In Figure 6-1, a simple space-division matrix is shown. At each end of the matrix are the wires (actual subscriber lines, line circuits, trunks, etc.) that are available for switching. They are represented as 1-N input lines and 1-N output lines. In

Table 6-2	**Switching Types**	**Functions**
Switching functions	Elementary Switching	
		Interconnection of input and output line circuits
		Control of communication traffic across line circuit groups
	Advanced Switching	
		Digit analysis
		Call record generation
		Route selection
		Fault detection
	Intermediate Switching	
		Monitor subscriber line circuits

Figure 6-1
An N × N space-
division matrix

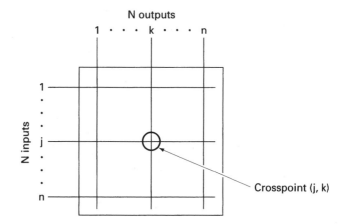

Figure 6-1
An N × N space-
division matrix

this example, the input J is connected to output K by closing the crosspoint (relay, contact, semiconductor gate, etc.): (J,K). Only one row of input can be connected to a column of output.

In networks utilizing space-switches, each call has its own physical path through the network.

Time-Division Switching

In time-division switches, the message paths are separated in time (hence, the name). A simple diagram of a time-division switch is shown in Figure 6-2, and it will serve as a simple way to explain this switching concept. In Figure 6-2, the subscriber units J1 through JN are in conversation with subscriber units K1 through KN by means of a time-division switch. The actual input (originating subscriber) and output (terminating subscriber) line circuits are opened and closed by individual switching devices, and are indicated as A1 through AN and B1 through BN. (The use of the input and output line circuits does not matter in this explanation. It is used to show some similarity between the space-division switch matrix example.) In the time-division matrix, the connection of subscribers takes place by controlling the operation of the selected switching devices A1–AN and B1–BN.

For digital time-division switches to operate, the incoming transmitted voice signals for every phone call must be in a digitized and encoded format. A T-carrier transmission system can provide the proper format to allow direct interconnection to a digital time-division switch without any additional conversion equipment.

Figure 6-2
Time-division switch
example

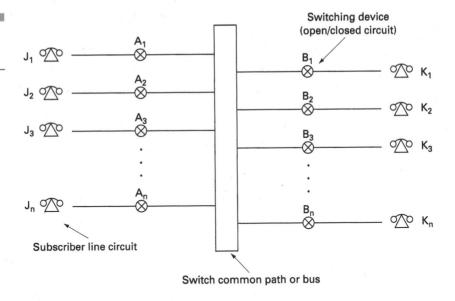

Switching device
(open/closed circuit)

Subscriber line circuit

Switch common path or bus

In summary, space-division switching involves switching actual circuit interconnections, whereas time-division switching involves switching actual digitized voice samples within the switch's matrix.

Packet-Switching

Voice communication systems utilize circuit switches to provide the switching function of voice line circuits. In data communication systems, packet-switches are used to perform the switching of data packets between the various nodes and computers in the network. Unlike the longer duration calls that require circuit-switching in a telephone network, packet-switching is better suited for short bursts such as transmissions of the data network. Packet-switching involves sorting data packets from a single line circuit and switching them to other circuits within the network. These sorting and switching functions are based upon an embedded network address within the data packet itself. An example of a packet-switch is a *Signal Transfer Point* (STP) in an SS7 data network.

Circuit-Switching Hierarchy

Within the North American *Public Switched Telephone Network* (PSTN), sometimes referred to as the "land-line telephone system," there are five classes of switches. These classes can be divided up into the central office class 5-level switch and the remaining tandem type switches of classes 1 through 4. The basic difference between these two categories of switches is that a class 5 local switch (also referred to as an end office) provides the ability to directly interconnect or interface to a subscriber's terminal equipment. A tandem-type switch will provide interconnection only to other switching equipment or systems. Thus, the local class 5 switch has the ability to terminate a call to one of its subscriber units; a tandem switch can route calls only to other destined nodes and never act as a final call delivery point (see Figure 6-3).

IP Networks

The IP network has gained wide usage with the introduction of the graphical interface for browsing the Internet. It is a true killer application (or a killer enabler). There is a mystique surrounding IP networks in that they are widely used, but just how they are used to foster communications is not known.

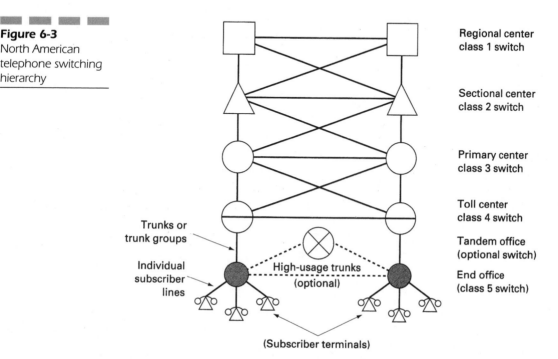

Figure 6-3
North American telephone switching hierarchy

Trunks or trunk groups

Individual subscriber lines

High-usage trunks (optional)

(Subscriber terminals)

Regional center class 1 switch

Sectional center class 2 switch

Primary center class 3 switch

Toll center class 4 switch

Tandem office (optional switch)

End office (class 5 switch)

An IP network can be a LAN, WAN, intranet, or the Internet. The vastness of what an IP network can really comprise has fostered some of the misunderstanding. Basically, an IP network is another protocol that is another enabler, allowing for more information to be transported. How and where it is transported can and does take on many forms.

Associated with the LAN or WAN is the use of an intranet, representing an internal network in which members of the same campus, corporation, or whatever share resources (at least some), but these resources are not shared to anyone outside. Typically, the LAN/WAN involves connecting a series of computers to a hub, which in turn might or might not be connected to a internal server, whether used for file sharing, the database, the web, or all of them.

There are numerous design books written with regards to IP networks, and each has its own specific slant—whether it is a service provider or is vendor-driven. Web sites such as http://www.cisco.com are excellent sources for information related to IP design and questions about routing and IP address designs.

However, the heart of an IP network is the fostering of an ASP because providing a large pipe by itself will not result in additional revenue over the customer's life cycle.

The biggest issue with LMDS is for the network engineer to ensure that there is sufficient bandwidth to support the various applications and services offered, and the correct CPU processing and memory to support the required additional services.

An LMDS operator can provide different levels of QoS for IP traffic using IPv4. The method utilized, depending on the infrastructure vendor chosen, is through use of VPNs. The QoS can primarily be guaranteed if the IP traffic remains on-net. But when it goes off-net, say to the Internet, the QoS can not be ensured, which is an obvious point.

Several general IP platform configurations are shown as follows for reference. The first, shown in Figure 6-4, illustrates a potential LMDS service

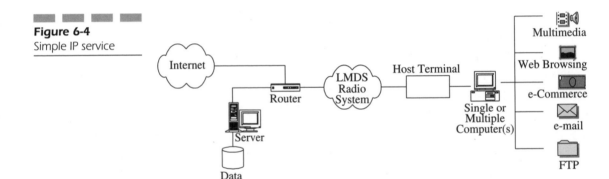

Figure 6-4
Simple IP service

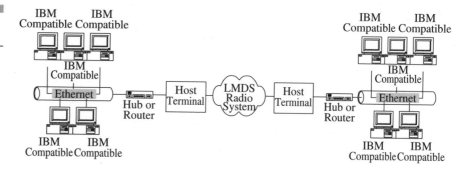

Figure 6-5
LAN-LAN/WAN

offering to either a residential user or small business that wants only web access or requests hosting capabilities as well as web access.

Figure 6-5 is an illustration of a LAN-LAN or WAN in which the LMDS system provides the connectivity between the offices. This configuration is based on a fundamental premise that the entire LAN-LAN or WAN system is on-net. If part of the system is off-net, the SLA with the provider that provides the missing link in the communication chain needs to be properly addressed. For the example, Internet access is not shown, but it could be if the service is required with the associated security requirements being put into place.

Figure 6-6 is an example of a WAN in which several telemarketing and Internet e-commerce companies utilize the LMDS system to provide connectivity with the warehouse for delivery and shipment. The companies also utilize the LMDS system for web hosting and e-commerce validation services.

Figure 6-7 is an example of the LMDS operator being a pipe provider and potentially adjunct service provider for a small or medium sized ISP. The pipe bandwidth can be anywhere from 128K to a full T1/E1. See Figure 6-8.

The previous figures indicate that there are numerous IP configurations available, from the simplest IP offering to an ASP value-added service. The delivery of IP data and the associated benefits of this exciting platform are available with all the LMDS platform providers. The key difference is how they deliver the IP data, either via multiple hops, a self-healing host terminal routers, or a simple IP host terminal; or via a T1/E1 host terminal whose alternative use is to provide IP connectivity.

IP Addressing

The issue of IP addressing is important to understand in any LMDS system design. No matter what the transport protocol or the infrastructure vendor

Figure 6-6
WAN

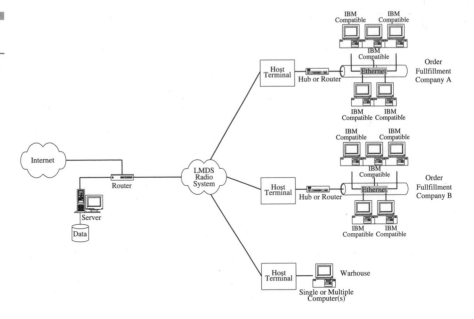

Figure 6-7
LMDS pipe provider

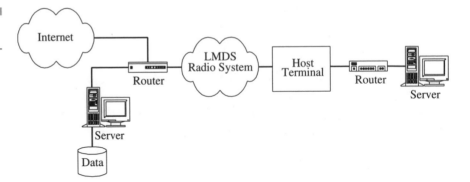

is for the LMDS system, the use of IP addresses is essential to the system's operation. It is imperative that the IP addresses used for the network be approached from the initial design phase to ensure a uniform growth that is logical and easy to maintain over the lifecycle of the system.

The use of IPv4 formatting is shown as follows. IPv6 or IPng is the next generation and allows for QoS functionality to be incorporated into the IP offering. However, the discussion will focus on IPv4 because it is the protocol today and has legacy transparency for IPv6.

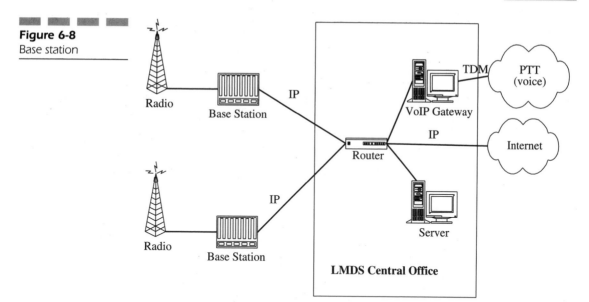

Figure 6-8
Base station

Every device that wants to communicate using IP needs to have an IP address associated with it. The addresses used for IP communication have the following general format.

Network Number	Host Number

Network Prefix	Host Number

There are, of course, public and private IP addresses. The public IP addresses enable devices to communicate using the Internet; private addresses are used for communication in a LAN/WAN intranet environment. An LMDS system will utilize both public and private IP addresses in order to facilitate the implementation of all its nodes.

Table 6-3 represents the valid range of public and private IP addresses that can be used. The private addresses will not be recognized on the public Internet system, and that is why they are used. It is also possible (and should be done) that an LMDS system can reuse private addresses within sections of its network, profound as this may sound. The concept is that the system can be segregated, and the segregation allows for the reusing of private IP addresses, ensuring a large supply of a seemingly limited resource.

Table 6-3

Public IP addresses

Network Address Class	Range
A (/8 prefix)	1.xxx.xxx.xxx through 126.xxx.xxx.xxx
B (/16 prefix)	128.0.xxx.xxx through 191.255.xxx.xxx
Class C (/24 prefix)	192.0.0.xxx through 223.255.255.xxx

Table 6-4

Private IP addresses

Private Network Address	Range
10/8 prefix	10.0.0.0 through 10.255.255.255
172.16/16 prefix	172.16.0.0 through 172.31.255.255
192.168/16 prefix	192.168.0.0 through 192.168.255.255

Table 6-5

Subnets

Mask	Effective Subnets	Effective Hosts
255.255.255.192	2	62
255.255.255.224	6	30
255.255.255.240	14	14
255.255.255.248	30	6
255.255.255.252	62	2

The public addresses are broken down into A, B, and C addresses, with their ranges shown as follows.

The private addresses that should be used are shown in Table 6-4.

To facilitate IP addressing, the use of subnetting further helps refine the addressing by extending the effective range of the IP address itself. The IP address and its subnet directly affect the number of subnets that can exist, and from those subnets the amount of hosts that can also be assigned to that subnet. See Table 6-5.

It is important to note that the IP addresses assigned to a particular subnet include not only the host IP addresses but also the network and broadcast address. For example, the 255.255.255.252 subnet, which has two hosts, requires a total of four IP addresses to be allocated to the subnet–two for the hosts, one for the network, and the other for the broadcast address. Obviously, as the amount of hosts increases with a valid subnet range, the more efficient the use of IP addresses becomes. For instance, the 255.255.255.192 subnet allows for 62 hosts and utilizes a total of 64 IP addresses.

Therefore, you might say, why not use the 255.255.255.255.192 subnet for everything? However, this would not be efficient either, so an IP address plan needs to be worked out in advance because it is extremely difficult to change once the system is being implemented or has been implemented.

Just what is the procedure for defining IP addresses and subnetting? The following rules apply when developing the IP plan for the system and the same rules are used for any LAN or ISP that is designed. There are four basic questions that help define the requirements:

1. How many subnets are needed presently?

2. How many subnets are needed in the future?

3. How many hosts on the largest subnet presently?

4. How many hosts on the largest subnet in the future?

Therefore, an IP plan can be formulated for the company's LMDS platforms. It is important to note that the IP plan should not only factor into the design the customers' needs, but also the LMDS operators' needs.

Specifically, at a minimum, the LMDS operators' needs will involve IP addresses for the following platforms. The platforms requiring IP addresses are constantly growing as more and more functionality for the devices is done through SNMP.

- Base Stations
- Radio Elements
- Microwave Point to Point
- Host Terminals (LMDS and Customer sides)
- Routers
- ATM Switches
- Work Stations
- Servers

The list can and will grow when you tally up all the devices within the network, both from a hardware and network management aspect. Many of the devices require multiple IP addresses in order to ensure their functionality of providing connectivity from point A to point B. It is extremely important that the plan follow a logical method.

Some examples of LMDS equipment require an IP plan that incorporates the entire system, not just pieces.

A suggested methodology is to do the following:

- List all the major components that are will be or could be used in the network over a 5–10 year period.
- Determine the maximum amount of these devices that could be added to the system over 5–10 years. A suggestion is to do this calculation by base station for the LMDS radio-related equipment.
- Determine the maximum amount of host terminals per sector that can be added.
- Determine the maximum amount of customers that can be connected to a host terminal.
- Determine the maximum amount of physical sectors that could be used per base station.
- Determine the maximum amount of LMDS radios and their type for each sector and base station.

The reason for the focus on the base stations and the host terminals is because these devices will have the greatest demand for IP addresses because of their sheer volume in the network. Also, if there are multiple customers per host terminal, the amount of IP addresses required will in all likelihood increase, depending on the type and number of ports on the Host Terminal.

For instance, the possibility of utilizing public addresses for all the devices is not practical, but some public addresses will be required. Therefore, the use of private IP addresses will be required to ensure that all the required devices have an associated IP address that has the correct network and subnet associated with the particular service.

A suggestion is to utilize the 172.16/16 through 172.31/16 addresses for the radio access portion, and the 192.168/16 for all the other devices. Obviously, this is not efficient either because it assigns entire blocks of numbers to specific devices without addressing the required usage. The associated subnets will also need to be established, with each IP address assigned or reserved. Depending on how the network is laid out, the reuse of IP addresses can take place if each city or region had its own domain name and NAT. The objective is to keep as much of the devices on the same network, so that the amount of retranslations and hops are kept to the extreme minimum.

The method that is suggested for arriving at the IP address assignment is to set the system up as a grid, usually the same grid that is developed by the RF engineering design process. The grid system then can be used for assigning IP address blocks to each grid, based on the device or service that can be offered there. See Figure 6-9.

For discussion purposes, the grid shown in the figure could have IP address 172.16.1.0 with a subnet of 255.255.255.0 assigned to sector A, and then the required IP addresses associated with it and the related subnets could now be associated with that sector. Sector B of cell 1 would then have IP address 172.16.2.0 and so forth. This method, although not the most efficient in the assignment method, would lend itself to operation simplicity.

Figure 6-9
IP grid

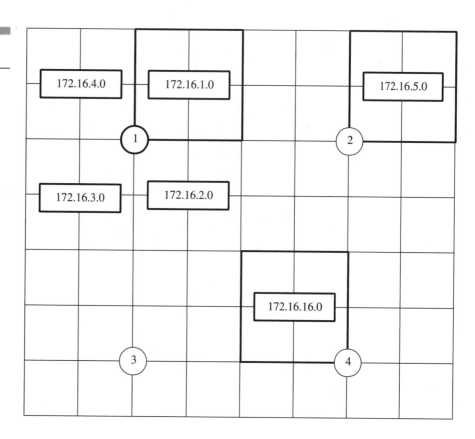

Using this logic for grid assignment, cell 2 would have 172.16.5.0 for sector 1, and so forth. But the logical question is what happens when you cell or sector split using this method? The answer is that you make the choice early in the system design to have each grid intersection have associated with it an IP address range or when cell splitting takes place. To accommodate this, a separate range of IP addresses following a similar address scheme is put in place, but beginning with a different starting IP address such as 172.25.1.0.

The IP assignment process could be designed so that the IP addresses are associated with a particular grid number instead of a sector of a particular cell. However, the IP address by sector will probably prove to be more user-friendly for Ops than having a grid and then another lookup cross reference table to the cell and sector.

Regarding microwave point to point systems, the IP address range of 172.31/16 could be used, but this depends on how and what they are connected to. And the rest of the devices can follow a similar logical assignment.

Naturally, your particular requirements will be different and the IP address method implemented. However, the concept presented has been beneficial and should prove useful.

Obtaining Public Internet Addresses

When designing and implementing an LMDS system, the use of public Internet IP addresses will be required as part of the systems operation. The public addresses will be needed to connect to the public Internet and also to have customers of the system utilize public addresses for operating their own ISP, which the LMDS system can be a transport media for. The amount of public IP addresses should not be that vast for the LMDS system because the use of private addresses will be the predominant pool of IP addresses to draw upon, which also includes the use of NAT.

But the LMDS system will require some public addresses for operating purposes. The quickest and most expedient method for obtaining public IP addresses is to obtain a block of IP addresses from another ISP because in all likelihood the system will connect to the Internet via another broadband service provider.

The recommended procedure is as follows:

1. Request IP addresses from an upstream provider.
2. Request IP addresses from the upstream provider's provider.
3. Request IP addresses from the following:

North America—ARIN, www.arin.net

Europe—RIPE NCC, www.ripe.net

Asia—APNIC, www.apnic.net

Each of these organizations has specific guidelines to follow for obtaining public IP addresses and can be readily extracted following the procedures outlined on the web pages.

Becoming an ISP

Every LMDS provider is also an ISP, or will be one. Just what is involved with becoming an ISP? The answer is very little; the system can be as simple as a Pentium computer running Linux or NT with a few modems. An ISP also does not need to provide dial-in capability; it can just host web pages, thus eliminating many of the logistical problems and capital per-subscriber costs.

The steps needed for becoming and ISP are as follows:

1. Obtain Internet access, 56k, 128k, 1M, etc. from a wholesale provider

2. Hardware and software to manage Internet communications *

3. Ethernet or dial-in lines (56k,ISDN) to connect Internet users to your Internet gateway

4. Personnel to manage system and customer inquiries and sales

Many companies are now providing virtual ISP service where they are acting as the wholesale side of the business. This significantly reduces the barrier to entry for starting up an ISP. It also reflects the low cost to entry and low margins that exist for just providing ISP connectivity.

The concept of being an ISP is no longer financially attractive if all you provide is dial-in service. The real market is in value-added services, such as e-commerce and other application-specific programs.

Therefore, the LMDS provider needs to offer to be the platform provider to small- and medium-sized businesses. In addition, the offered service to existing ISPs can also be made on the premise that what is offered may be larger for the same fee as they are paying from the PTT or CLEC, and that more services are available as they can grow, such as becoming a virtual ISP.

---

* Becoming an ISP does not necessitate the need to have physical equipment.

ATM Platforms

Although ATM transport has been utilized by the PTT for some time, with the introduction of LMDS and the need to transport data as well as possibly voice, the desire to utilize a transport media that can handle both continuous traffic and bursty traffic is essential. With the continuous push from more IP-related services, the network designer for a LMDS system is faced with the dilemma of ensuring different grades of service and QoS, while at the same time not overdimensioning the various pipes within the network so as to reduce recurring facility costs.

Entire books are written on the design and functionality of ATM switching. It not my intent to rewrite those excellent books, which can be found in the reference portion of this chapter. Instead, the objective is to show how the ATM platforms fit into the LMDS service platforms, depending of course on the services and functions that are offered by the system.

ATM is, by default, a connection-oriented transport method that is excellent for transporting all types of information content, specifically data traffic and effectively voice services. The ATM switch performs two major functions routing, self and label—in addition to header translation.

The ATM routing types are *sequential routing* (SR), *random alternative routing* (RAR), *least loaded routing* (LLR), and *minimum cost routing* (MCR).

In an ATM switch, the connections are referred to as *virtual circuits* (VCs), but the VC is just a container for tributary paths called *virtual paths* (VPs). There can be multiple VPs within a virtual circuit, as shown in Figure 6-10.

Figure 6-10
Virtual path

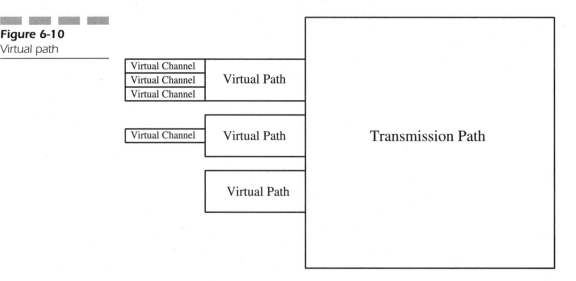

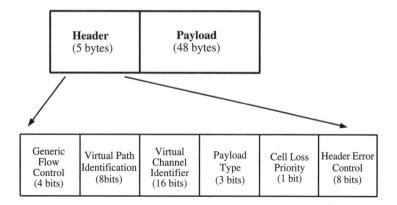

Figure 6-11
ATM cell format

ATM is able to provide greater throughput in the network because it utilizes fixed-length packets called cells. A cell consists of 53 bytes, of which 48 bytes are for data and five bytes are for the header. Figure 6-11 illustrates the makeup of an ATM cell and the various components that comprise the ATM cell itself.

Referencing the figure, the various fields have specific functions. More specifically

■ Generic Flow Control (*GFC*)   This field's purpose as it implies by its name is to control the traffic flow on ATM connections. This field is also used for CBR services for the purpose of controlling jitter.

■ Virtual Path Identifier (*VPI*)   This field's basic function is to identify the VPI used for connecting the *Virtual Channel Identifier* (VCI).

■ Payload Type (*PT*)   This field provides the type of data contained within the cell. Types of information the PT will identify include user data, signaling data, congestion information, and maintenance data.

■ Cell Loss Priority (*CLP*)   This is the field that identifies whether the cell should be discarded during congestion.

■ Header Error Control (*HEC*)   This field's purpose is to provide the error correction.

The ATM transport method allows for different service classes, and those classes are directly related to the AAL layer. There are five AALs, and each is meant to transport a particular type of service. The type of service for each of the AALs and the connection type are listed in Table 6-6 and Table 6-7.

This table is interesting, but without relating the specific AAL levels to possible LMDS service offerings, it has little value for the LMDS operator. Therefore, the following is meant to help further refine the table:

Table 6-6

AAL

Service Class	AAL	Bit Rate	Timing Relationship between Source and Destination	Connection Type	Applications
A (1)	1	Constant	Required	Connection-oriented	CES CBR Video
B (2)	2	Variable	Required	Connection-oriented	Rt-VBR Audio/ Video (Multimedia)
C (3)	5 (3)	Variable	Not Required	Connection-oriented	Frame Relay FTP
D (4)	5 (4)	Variable	Not Required	Not Connection-oriented (that is, connection-less)	IP,SMDS

Table 6-7

ATM class/type

Class/ Type	Bandwidth Guarantee	Real-Time Traffic	Bursty Traffic	Congestion Feedback
CBR Constant Bit Rate	Yes	Yes	No	No
RT-VBR Variable Bit Rate, Real Time	Yes	Yes	No	No
NRT-VBR Variable Bit Rate, Non-Real Time	Yes	No	Yes	No
ABR Available Bit Rate	Yes/No	No	Yes	Yes
UBR Unspecified Bit Rate	No	No	Yes	No

Constant Bit Rate (CBR). This supports applications that require a fixed data rate being always provided. Examples of where CBR is applied involve TDM circuits, and voice and leased lines such as T1/E1.

Real-Time Variable Bit Rate (rt-VBR).   This service supports applications that require real time data flow control. An examples of where rt-VBR is used involves video conferencing. The main difference between CBR and rt-VBR is that rt-VBR has tighter timing controls than those used for CBR.

Non-Real Time Variable Bit Rate (nrt-VBR).   This service supports applications, which by its name do not require exact timing between source and destination. Some applications for nrt-VBR involve email and multimedia applications.

Available Bit Rate (ABR).   This service is really an enhancement to UBR in that a minimum and peak cell rate can be defined. ABR has priority over UBR traffic. Effectively, ABR is a managed best effort service.

Unspecified Bit Rate (UBR).   This service is also referred to a best effort service. It supports applications that can tolerate variable delays between source and destination plus the possibility of cell loss. UBR traffic uses the bandwidth that is left over after CBR and VBR services have taken their share of the pipe. The UBR service allows more utilization of the ATM network by passing UBR traffic at different rates between CBR and VBR allocations.

The next logical question that should arise is what are the performance parameters that should be used for evaluating the health and well being of the ATM network or its components? Table 6-8 contains the key ATM parameters that measure the performance of the system.

ATM Performance Parameters

Table 6-8 lists some of the key performance parameters associated with an ATM system.

ATM Networks

Just what is an ATM network? To start with, an ATM network can consist of a single ATM switch or multiple ATM switches. When you want to leave the ATM media world, it will require converting to the appropriate protocol such as TDM or IP, which is usually done with a ATM edge switch.

Table 6-8

ATM performance
parameters

Parameter	Measures	Calculated
Cell Error Ratio	Accuracy	Errored Cells/(successfully transferred cells + errored cells)
Severely Errored Cell Block Ratio	Accuracy	Severely error cell blocks/total transmitted cell blocks
Cell Loss Ratio	Dependability	Lost Cells/Total Transmitted Cells
Cell Misinsertion Rate	Accuracy	Misinserted cells/time interval
Cell Transfer Delay	Speed	Elapsed time for a cell between the source and destination
Mean Cell Transfer Delay	Speed	Average of a defined number of cell transfer delay estimates
Cell Delay Variation (CDV)	Speed	Variability in the pattern of cell arrival events to a single point versus the negotiated contract rate (SLA)

Figure 6-12
ATM network
interfaces

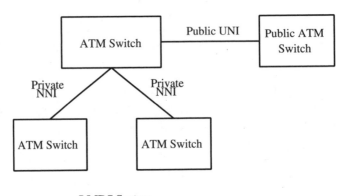

LMDS System

ATM systems support two general types of interfaces:

User Network Interfaces (UNI)

Network Network Interface (NNI)

Figure 6-12 represents the location of the different interfaces. Basically, the private NNI is when you connect to ATM switches within your own network; the public UNI is when you connect to the public ATM switch or cloud.

There are also several types of NNIs for ATM networks:

Private Network to Network Interface (PNNI)

Broadband Inter-Carrier Interface (B-ICI)

Broadband ISDN Services User Part (B-ISUP)

Interim Interswitch Signaling Protocol (IISP)

ATM routing is centered on circuits of two types: VPIs and VCIs. Comprising these is the use of SVCs and PVCs.

Switched Virtual Connection (SVC) is a connection that is set up dynamically, based on the need for the connection. Specifically, every time service is requested, the path taken from the source to the destination can change based on the resources available. SVCs make best use of the network's facilities by increasing its utilization.

Permanant Virtual Connection (PVC) is a predefined route that is programmed into the ATM switch via a craft person. In the diagram from Figure 6-13, all of the paths could be considered PVCs as along as the VPI/VCI and ports remain the same.

Figure 6-13 is an illustration of the routing functions that are associated with an ATM switch. Naturally, there is more then the simple port mapping illustrated in Figure 6-13, but the diagram helps facilitate the issue of VPI and VCI mapping. Also see Table 6-9.

The following are a few simple ATM network layouts that are possible and probable when utilizing ATM platforms in conjunction with LMDS.

Figure 6-14 is a simple example of an ATM system that involves a core ATM backbone for an LMDS operator. The ATM core backbone is rarely utilized from the start, but is instead a planned event where the need for larger capacity ATM switching platforms becomes more evident as traffic grows. Traffic growth by itself is not the sole reason for expansion of an ATM backbone. The key driver for the ATM backbone is the volume of bursty traffic requiring better management of the bandwidth that is not readily available with a TDM backbone. An IP backbone can handle bursty traffic, but with IPv4, the QoS is not available, thereby missing out on many of the opportunities that broadband data can provide to the end user.

Figure 6-15 shows the situation where different portions of the LMDS system are connected to a core ATM network. The edge ATM switch not

Figure 6-13
ATM switch

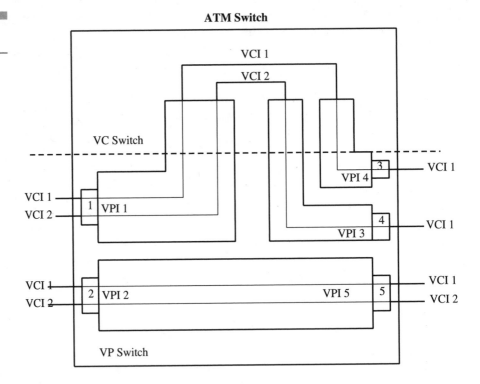

Table 6-9

ATM switch
mapping

Input			Output		
Port	VPI	VCI	Port	VPI	VCI
1	1	1	3	4	1
1	1	2	4	3	1
2	2	1	5	5	1
2	2	2	5	5	2
5	5	1	2	2	2
5	5	2	2	2	2
3	4	1	1	1	1
4	3	1	1	1	2

Figure 6-14
ATM core network

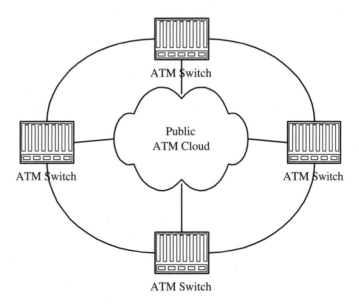

Figure 6-15
LMDS ATM backbone
with edge switch

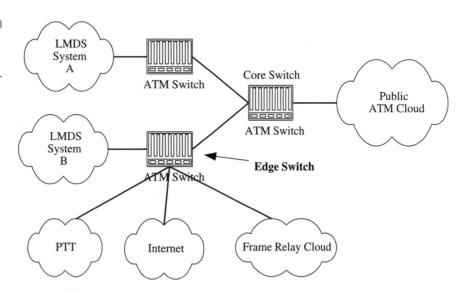

only is the main ATM switch for the regional LMDS system, but also provides connectivity to the Internet, Frame Relay cloud, and the PTT. The edge switch has the capability of converting from ATM format to the other required formats via circuit emulation.

Figure 6-16 is a depiction of an ATM edge switch being located with or near an LMDS base station. The advantage with this scenario is reduced facility costs at initial startup due to smaller bandwidth pipes at system or site inception times. The LMDS central office then performs the facility concentration that is needed or desired to provide lower interconnect costs through volume purchase agreements.

Most of the connection between the LMDS base station and the central office is handled initially by the PTT due to the amount of POPs they have. However, it is possible to utilize a CLEC and coordinate the RF design with the POP locations, provided there is a good technical match.

ATM Design Aspects

The design aspects for ATM switch designs focus on several key attributes, which are listed as follows:

■ Traffic characteristics

 ▪ *Burstiness*   Commonly used to measure how infrequently the traffic volume and rate is between the source and destination
 Burstiness = Peak Rate/Average Rate

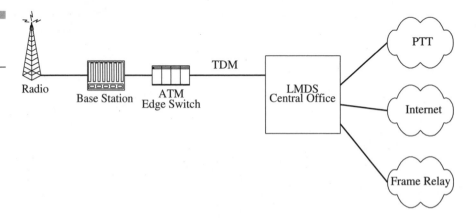

Figure 6-16
LMDS base station
ATM edge switch

- Traffic Delay Tolerance
- Response Time
- Capacity Throughput

- Cell Delay

- Cell Loss

- Congestion

- Ports Available

Traffic characteristics are important to factor in because an ATM switch is designed to handle CBR, vBR, and UBR traffic. The type of traffic and QoS associated with each, and the burstiness of the traffic all factor into the overall throughput for the switch. Typical design limits put the desired throughput for the design to be 70% of the allowable limit of the ATM switching platform. Throughput is not the same as the amount of ports.

The other topics, such as cell delay, factor into the ability to deliver particular time-sensitive traffic. The cell delay is important for the individual switch, but is really related to the overall network design, and in particular to the path the cell has to transverse from start to finish.

Cell loss is important for many protocols that do not have error correction. Additionally, if there is a lot of cell loss, the link loads increase due to retransmissions which has a direct effect on the throughput and cell delay factors.

Congestion by itself is an important issue to avoid for an ATM switch design. During high congestion periods, heavy cell loss can occur by design, resulting in more congestion due to the amount of retransmission that takes place.

The port issue is important when connecting to different ATM networks and providing circuit emulation to a TDM platform. Many times, the amount of ports is the driving issue for ATM edge switches and not the other design parameters. The ports for an ATM switch can run at 100% utilization. However, in the initial design phase with unknown traffic projections, the desired level is 70% of the estimated load for which every ATM switch platform is chosen, at the design end point. This is of course based on the premise that the ATM platform chosen has the available card slots to accommodate this potential growth.

Facility Sizes

Table 6-10 is a brief chart, detailing the different facilities that a LMDS operator may have to interface to and the associated bandwidth each is associated with.

Table 6-11 is a quick reference of the similarities between a SONET and SDH fiber system.

Table 6-10

Facility sizes

Signal Level	Carrier System	Number of DS1 Systems	Mbits/sec
DS0	DS0	1/24	0.064
DS1	T1	1	1.544
DS1C	T1C	2	3.152
DS2	T2	4	6.312
DS3	T3	28	44.736
DS4	T4	168	274.76
OC1	OC1	28	51.84
OC3	OC3	84	155.52
OC12	OC12	336	622.08
OC48	OC48	1344	2488.32

Table 6-11

Fiber sizes

Signal Type	Mbps	SDH
STS-1/OC1	51.84	—
STS-3/OC-3	155.52	STM-1
STS-12/OC-12	622.08	STM-4
STS-24/OC-24	1244.16	STM-8
STS-48/OC-48	2488.32	STM-16

LMDS Central Office

There is no single correct layout for an LMDS central office, except that it needs to provide the necessary space and functionality for the various service platforms that need to be installed at that location. The Central Office should be designed to accommodate growth at that location for a period of 5 years before an alternative relief site is sought.

A sample LMDS central office, as shown in Figure 6-17, includes functionality for ATM, TDM, and IP platforms. A class 5 switch is included with the layout, but will probably not be required for a variety of reasons, possibly because you will not sell traditional voice services via LMDS.

Demand Estimation

Determining the demand that the system will need to transport and the associated bandwidth that is needed can be reasonably calculated given marketing forecast data. The marketing forecast data is important for both a new and existing system because it is the best way to estimate the

Figure 6-17
LMDS central office

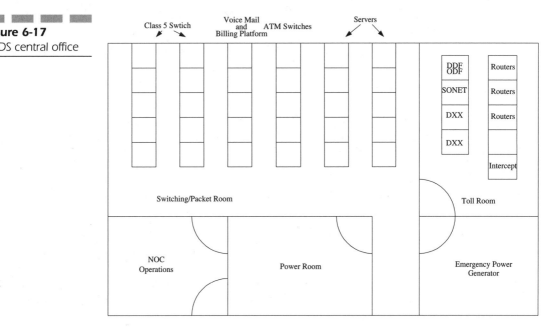

amount of traffic that a system will need to transport and the media type associated with the traffic load.

The key difference between a new and an existing system design lies in the issue of having a baseline from which to begin the forecast from. If the system is new, there is no baseline. However, if the system is in operation, there should be some traffic being carried by the system and therefore the marketing forecast is an addition to the current traffic being carried or designed.

$$Design\ Traffic\ =\ Current\ Carried\ Traffic\ +\ Forecasted\ Traffic$$

The key point is always trying to determine how far in advance the study needs to be done and the frequency of the study. The recommendation is that the study should not be done in detail for more than two years. The first year should be broken down by quarters and the second year should be broken down in six-month intervals. The traffic estimation will need to be revisited on a three- or six-month basis for the life of the system to ensure that proper dimensioning is taking place and to account for the hot spots or lack of take rates estimated.

The traffic shown in Table 6-12 has multiple services included. Naturally, if the service is not offered by the LMDS operator, it can be excluded from the following list. However, it is always easier to take away than to add.

A design assumption is that a total of 50 base stations will be deployed with this network.

For the forecasting, there are several key elements that the network engineer needs to factor into the forecast and design:

- Types of services
- Volume of traffic for each service
- Platform requirements and growth
- Connectivity between base station and central office or concentration node
- Connectivity between concentration node and the various transport providers (PTT, CLEC, IP)
- Timeframe of study
- Current traffic utilization

Besides the platforms that are associated with the LMDS system, the various interconnect pipes need to be designed. Figure 6-18 represents some of the interconnect issues that need to be designed. The figure assumes that there is one provider for each service platform. In reality, however, there

Table 6-12

Total system BHT

Grid Number	Type Type			Number Sub	Oversub- scription Factor	Total Total Mbps
AAL1	**VPN/WAN**	CIR	Burst			
	T1/E1	1.5M	NA	150	0	225
	T1/E1 (IP)	1M	1.500	500	10	525
	Frame Relay 128k (SVC)	.064	.128	200	10	14.08
	Frame Relay 512k (PVC)	.512	NA	100	0	5.12
	ATM (PVC)	1M	1.5M	25	5	27.5
	X.25	0.064	NA	50	0	3.2
AAL1	**Circuit Switched**					
	BRI (1B+D)	0.064	NA	500	0	32
	BRI (2B+D)	0.064	0.128	250	10	17.6
	PRI (10-24)	0.64	1.536	750	5	710.4
	POTS	0.064	NA	100	0	6.4
AAL5						
	IP (Always On)	.256	.512	1000	30	264.53

could be multiple service providers for each type of service offered. For this example, it is assumed that the traffic is 100% off-net for delivery. If the traffic were not 100% off-net, then it would be necessary to include the additional capacity for delivering the service over the LMDS radio systems links. I excluded this fact for simplification at this time. See Table 6-13.

The size of the pipes from the previous forecast involves Table 6-14.

Depending on the service providers, a OC-48 would be required to support this facility's bandwidth requirements for the services only. The pipe sizing shown does not reflect different CIRs, their associated SLAs, or the connection between the base stations and central office.

Figure 6-18
Interconnect

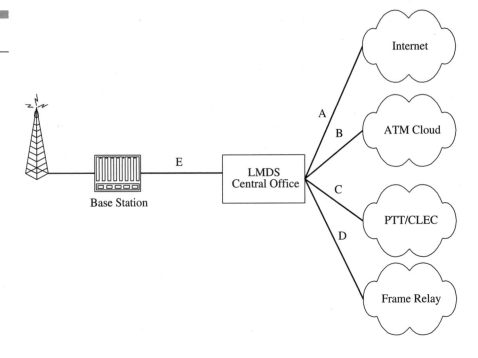

Figure 6-18
Interconnect

AAL1		VPN/WAN	Mbps	DS1s
T1/E1			225	146
T1/E1 (IP)			525	340
Frame Relay			19.2	13
ATM (PVC)			27.5	18
X.25			3.2	3
AAL1		**Circuit-Switched**		
BRI (1B+D)			49.6	
PRI (10–24)			710.4	
POTS			6.4	
Total			748.8	485
AAL5				
IP (Always On)			264.53	172

Table 6-13

Traffic

Table 6-14

Pipe sizes

Connection	Service	Pipe
A	IP	OC-12
B	ATM	T3
C	Voice + Leased Lines	OC-24
D	Frame Relay	T3

Regarding the base station to central office link dimensioning, the following list gives a quick assessment of the situation:

50 base stations

Traffic uniformly distributed by service and type

1177 Mbps total system load

Therefore, this equates to 23.54 Mbps for each base station, which is 16 T1s or 12 E1s. Either way, a T3/E3 would need to be secured for each of the links from a base station to the central office. The E3 would provide about 25% headroom, whereas the T3 would provide about 45% headroom.

Therefore, with 50 base stations, and each having a T3 associated with each one, an OC-48 would be required at the central office just to handle the LMDS systems backbone. When combing the backbone with the services, an OC-96 seems more appropriate for location, which is anything but trivial for size.

References

Azzam, Albert A. *High Speed Cable Modems*. New York, NY: McGraw-Hill, 1997.

Bates, Regis J. and Donald W. Gregory. *Voice and Data Communications Handbook*, 3rd ed. New York, NY: McGraw-Hill, 2000.

Black, Uyless D. *TCP/IP and Related Protocols*, 3rd ed. New York, NY: McGraw-Hill, 1997.

Goralski, Walter J. *ADSL*. New York, NY: McGraw-Hill, 1998.

Guizani, Mohsen and Ammar Rayes. *Designing ATM Switching Networks*. New York, NY: McGraw-Hill, 1998.

McDysan, David E. and Darren L. Spohn. *ATM Theory and Application*, Signature ed. New York, NY: McGraw-Hill, 1998.

Russell, Travis. *Signaling System #7*, 3rd ed. New York, NY: McGraw-Hill, 2000.

Smith, Clint and Curt Gervelis. *Cellular System Design and Optimization*. New York, NY: McGraw-Hill, 1996.

Winch, Robert G. *Telecommunication Transmission Systems*, 2nd ed. New York, NY: McGraw-Hill, 1998.

Host Terminal

Introduction

The host terminal is the device used by the LMDS system to provide customer access to the broadband system. The host terminal will usually consist of a radio interface device that converts the RF energy into a usable format and vice versa. The host terminal consists of an *Outdoor Unit* (ODU), usually the radio and an antenna. The ODU radio and antenna can be an integrated unit or consist of two separate devices. The ODU is then connected to an *Indoor Unit* (IDU) that provides the physical or medium conversion from RF to electrical, enabling the customer to utilize the RF network provided.

Figure 7-1 is a generic example of the components related to a host terminal.

From the customer's aspect, the use of a radio system that is the technology enabler should be a non-issue. If the system is designed properly, all that should matter to the customer is that the necessary interfaces are available to support the type of service offered by the LMDS operator and desired by the customer, a required match.

The interface that the LMDS system has with respect to the customer is referred to as the *Customer Interface* (CI). However, what the interface is differs from business, infrastructure vendor, and of course services offered by the LMDS company. The interface that is being referred to in this chapter is not the company vision or the pre- and post-sales organization, but

Figure 7-1
Host terminal

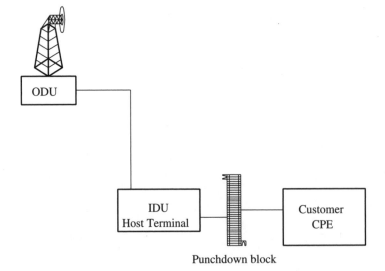

Punchdown block

the interface as it pertains to the host terminal, which is placed on the customer's premise itself.

Just what is a customer interface? The answer is almost as varied as the technology platforms that are available. From the LMDS company's perspective, the interface should be consistent and inexpensive to not only implement but also to maintain and possibly expand. From the customers aspect the interface needs to provide the necessary hooks that they require without requiring any infrastructure changes to their premise equipment. Obviously, having an interface that presents uniformity and operations simplicity has a great appeal for any LMDS operator. From the customer's aspect, the interface should be such that in order to take advantage of the services, they do not have to replace any existing CPE but just change the service provider as the demarcation point.

However, providing a flexible interface that fits all situations and presents the minimal operation issues results in a configuration that may be too expensive to implement. On the other hand, the interface could also be so rigid that when the customer wants to expand or add to the service portfolio, the host terminal interface to the customer cannot be easily expanded.

One of the key concepts that all LMDS operators are struggling with, or have struggled with, is the man in the truck problem. You do not want to deploy that man in the truck each time a service order is issued for service expansion, and hopefully not in the other direction. At the same time, over-provisioning by too much can also present capital constraints that are not economically viable to pursue.

Another aspect with the man in the truck issue lies in operations personnel costs, OpEx, and the cost of the host terminals themselves. As more and more customers are added to the system the idea is to reduce (that is, spread) your capital costs across the widest or largest amount of customers, thereby reducing the cost per customer in capital, which also equates to cost per port or service delivery.

Along with the man in the truck is the decision to outsource this function or internalize it. The decision to internalize or externalize the man in the truck and other adjunct services that are required by the customer is driven by the following: marketing/sales objectives, OpEx objectives, VARs and JVs that are pursued, the skill sets and personnel in the organization, and the cost per customer/port required for economic viability at which the LMDS system needs to perform (which is driven by cash flow and capital constraints). From this brief list of issues (there are more, of course), the interface from the host terminal to the customer premise equipment is not straightforward and takes on multiple facets that will differ significantly, based on the services offered, the infrastructure of the host terminal, and the market conditions.

Demarcation Decision

The demarcation decision is not simple and cannot be relegated to the purview of a PowerPoint slide. There is no one answer for demarcation because it can and does involve a multitude of issues. The demarcation needs to address the physical, electrical, practical, and policies implemented, which should be approached on a uniform basis where the combining of the parts provides the level of service that ensures customer satisfaction while at the same time ensuring the economic survivability of the company. Although this seems simple and rather straightforward, this is not as simple as it may appear.

The demarcation for an LMDS company can and will differ, based on the type of customer and economic potential that are at the other end of the host terminal. More specifically, if the service offering is high-speed IP traffic to residential homes, the level of support is significantly different than that demanded by a company that requires a CIR of 1M with burst capability of 1.5M or 2M for IP traffic.

Addressing the demarcation issue in more depth, the main topics that should be considered in establishing the interface addresses the previously mentioned topics. See Figure 7-2.

- Physical
- Electrical (physical and logical)
- Practical (support)
- Policy

Figure 7-2
Demarcation

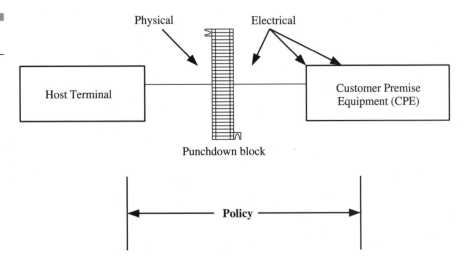

Regarding the physical demarcation point, this addresses the physical aspects. For instance, if the host terminal is physically installed in the customer's facility, the decision of how you connect to their CPE needs to be defined in advance to avoid the vast array of issues that can come to materialize. For instance, do you provide a punchdown block and terminate to the block or do you physically connect to the customers CPE via a direct cable?

The issue is complex, depending on the service type offered. IP traffic, using a UPT CAT5 cable for instance, can allow for up to 100 meters between the host terminal and the customers CPE. However, if the service also provides Frame Relay via an X.21 connector, the issue is distance from the host terminal to the FRAD itself.

This issue is not that important if there is one customer and the host terminal resides in the customer's telco room or wiring closet. However, the physical issues become more perplexing with a multi-occupancy building and the host terminal interfaces available, coupled with the services offered.

The electrical demarcation is another aspect to factor into the decision process. For instance, when delivering IP traffic to a residential service, the electrical demarcation could be the host terminal itself or the NIC card interface into the computer (cable in-between) or the entire computer, where it is remotely managed and configured by the LMDS system.

The electrical demarcation point for a commercial service such as T1/E1 may prove to be tricky because the interface should be the punchdown block that is provided. But when a problem occurs, the need to isolate the problem will need to exist (that is, separate the host terminal from the customer premise CSU/DSU). But if there is no intelligence with the cross connect panel, the isolation method is not complete because the cabling wiring between the host terminal and the CPE could be at fault. This could result in the man in the truck visiting the location for troubleshooting, while all the time the system is not operational and is in possible violation of the SLA.

The policy that is in place for defining the demarcation point between the host terminal and the CPE is critical. Specifically, what support do you plan on offering? Offering complete 100% support when delivering just IP traffic could easily involve fixing computer configuration, software and hardware problems that have nothing to do with the LMDS service offering. You cannot just turn a blind eye to the customer who has a problem, but this does not mean that you just offer the additional support by charging them $250 for a service that they pay $35 per month for. However, you cannot expect to include this expense as part of the operating budget.

The LMDS operator needs to determine when to stretch the policy for support to ensure that the customer problem is resolved, even if the problem is not under your direct responsibility but is impacting the customer's ability to utilize your network.

From the practical aspect, the support team, customer care, and technical community need to have scripts defining troubleshooting techniques, in addition to incorporating policy issues with the scripts where escalation within the customer care organization may need to be brought to bear on the situation. For instance, for corporate IP customers, a number of support calls for non-service related calls may be allowed and included on the customer's profile screen. They indicate the amount of credits they have been granted via sales that should be directly related to the services taken and the profile of the customer. The profile specifically relates to the issue that a service may be ordered, Frame Relay 256K, for example. But because the company is a major data center and you are currently an alternative service provider being explored for potential future use, the credit level should be different as well as the account being flagged for special attention by senior customer care and technical resources.

Customer Interface (CI) Types

There are numerous types of *customer interface* (CI) types that a LMDS company can apply or have available for connecting a customer to their system. The CI interface needs to address the demarcation issues, physical, electrical, and logical. Of course, these need to be consistent with the company, as mentioned before. However, the choices of what is available are many, even for a single service type offering (that is, a high-speed Internet access IP, in which the LMDS host terminal hooks directly to the customer's computer).

The configuration CI type is directly dependent upon the infrastructure vendor chosen for the platform and the service offering by the LMDS company.

Following are some examples of CI types that are possible. The configurations shown can, of course, be modified based on the particular services offered. All of the configurations shown have excluded the radio portion of the host terminal.

The first set of configurations address IP services only when the host terminal has either a single or multiple 100/10 BaseT connector that enables the customer CPE to connect either directly to the host terminal or through a punchdown block. The normal method for this configuration is to allow the customer CPE to connect directly to the host terminal because this con-

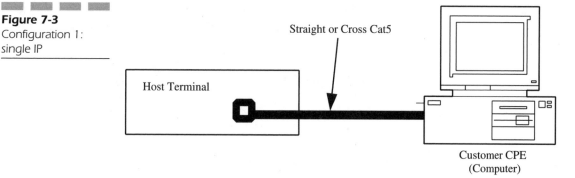

Figure 7-3
Configuration 1:
single IP

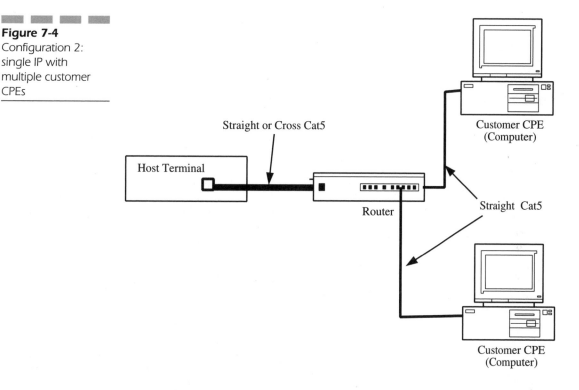

Figure 7-4
Configuration 2:
single IP with
multiple customer
CPEs

figuration represented is a 1:1 relationship with the host terminal to the customer (no multiplier).

Figure 7-4 represents a logical extension of Figure 7-3, in which there are multiple computers that are on a customer's location where the desire is to provide broadband access to all the units. Figure 7-4 shows a router for this configuration instead of a hub or LAN switch. This reduces the amount of

IP traffic intended to be between the CPE, and not needed to be sent over the air—which would reduce the overall bandwidth available for other customers to utilize if overbooking is used.

Figure 7-5 represents the situation in which the host terminal has multiple IP ports available for use in delivering service to a customer or customers. The figure shows multiple CPEs connected to the terminal, which are implied to be from the same customer. In reality, however, the CPE can be from the same or different customers, provided that the interface is a layer 3 device.

Figure 7-6 represents a host terminal that is designed to support leased line replacement service. The figure represents a total of four interface ports, but the number of ports is directly dependent upon the spectrum available, besides the technology transport media.

Figure 7-5
Configuration 3: multiple IP interfaces

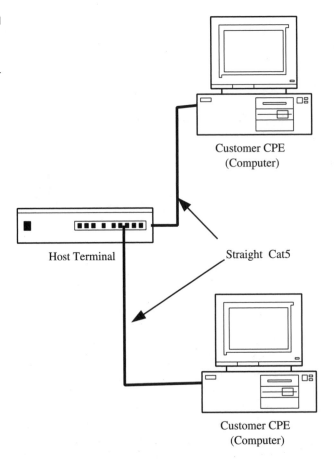

Customer CPE
(Computer)

Host Terminal

Straight Cat5

Customer CPE
(Computer)

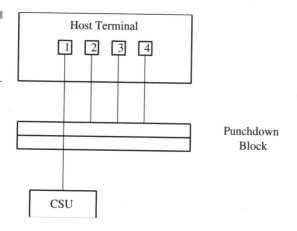

Figure 7-6
Configuration 4:
leased line
replacement

The leased line replacement service, depending on the host terminal, could also be used to support clear channel T1/E1 or fractional T1/E1 or $n*64$ Frame Relay, ISDN (PRI), etc. That is, provided that the interface supports both the physical and software port requirements.

The next figures represent the situation when voice services are desired to be delivered in conjunction with the leased line replacement services. Figure 7-7 represents an extension of the leased line replacement; Figure 7-8 shows what is possible when the interfaces are derived through the use of personality modules.

The obvious drawback in the situation shown in Figure 7-7 is the use of an additional interface platform to achieve the delivery services. For instance, to deliver ISDN BRI service or POTS, the use of a *Multiplexer DeMultiplexer* (MUX) is required. The ideal situation, currently in use with multiple service platforms available on the market, is to utilize personality modules that plug into the host terminal, enabling different services from the same fundamental platform.

The telephony services represented in Figure 7-8 could be analog POTS or ISDN BRI.

Figure 7-9 represents a common LMDS host terminal in which leased line services and their associated derivatives are provided in addition to one or multiple 100/10 Base T ports. The configuration shown in Figure 7-9 will allow for reasonable flexibility with integrating with the varying needs to the LMDS customers.

All the examples shown illustrate some of the possible variants that are available for use. However, when an adjunct platform is utilized in conjunction with a LMDS host terminal, the service offering potential increases, as well as the cost of the delivery equipment itself and operational complexity.

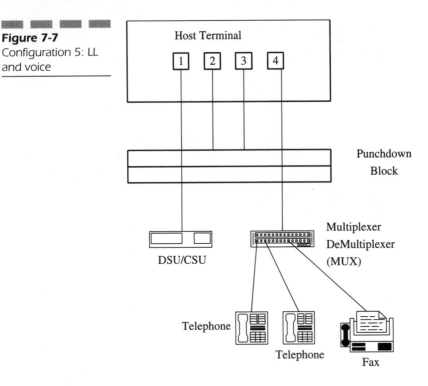

Figure 7-7
Configuration 5: LL and voice

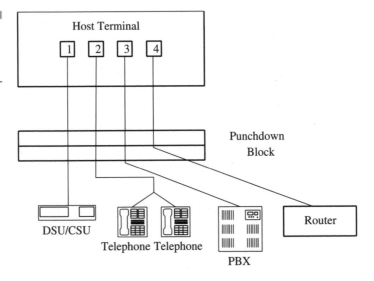

Figure 7-8
Configuration 6: personality module host terminal

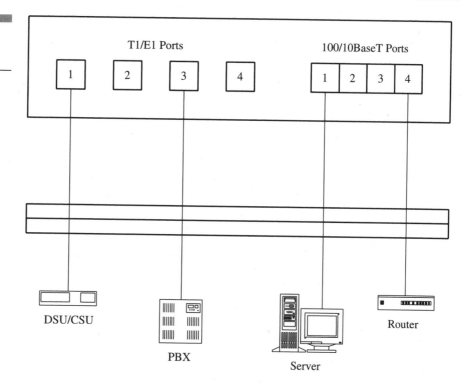

Figure 7-9
Configuration 7:
lease line and IP

T1/E1 Ports

100/10BaseT Ports

1 2 3 4

1 2 3 4

DSU/CSU

PBX

Server

Router

Services

The services that are available from the host terminal from an LMDS operator are directly dependent on the technology platform chosen as well as the services being sold. There are a many types of services that are enabled by the use of broadband brought about by LMDS.

The services that can be effectively delivered via the host terminal need to match the service level agreements that will be offered. For instance, if the SLA that is to be offered from the LMDS operator involves a *Committed Information Rate* (CIR) of 1Mbps of IP traffic with bursting capability of 1.5Mbps, the infrastructure utilized cannot deliver IP on a best-effort-only basis. This is an obvious issue, but one that needs to be addressed with every SLA offered to the customer base.

The services offered will also have a direct impact on the spectrum utilized. For instance, if leased line services are all that are offered, it is a constant bit rate service and effectively will not allow for overbooking for the spectrum. The overbooking or lack of overbooking of the spectrum will have a direct impact on the cost per port or per customer that is evident in an LMDS system.

Services can be defined as both transport and value-added. The differentiation is important but often they are not separated in the talks and decisions that are being put forth by marketing and sales. Transport services relate to the physical port interface services that are available from the host terminal. Provided the ports on the host terminal can support $n*64$ Frame Relay or $n*64$ T1/E1 services. This is different from offering a Frame Relay service from location A to location B because location B may or may not be on-net. The Frame Relay service of providing a PVC or SVC from point A to point B is a value-added service because the host terminal only addresses part of the service offering.

Another example of the difference between transport and value-added lies in the use of IP services. The transport service involves the ability to transport IP-related information. However, the value-added service is providing connectivity to the Internet or providing a LAN-to-LAN connection, which can be on-net or off-net and with the associated SLA.

Table 7-1 is meant to foster the discussion that the marketing and sales department needs to have with the technical organization within the LMDS company. The table is meant to show how the configurations listed in the previous section relate to services being offered.

The table makes many assumptions regarding the host terminal interfaces, such as the ability to deliver fraction services in addition to whole

Table 7-1

Possible edge terminal configurations

Services	Configuration	Comments
IP	1, 2, 3, 6, and 7	
T1/E1 Leased Line Replacement	4, 5, 6, and 7	
N*64 DSO (T1/E1)	4, 5, 6, and 7	
N*64 Frame Relay	4, 5, 6, and 7	
X.25	4, 5, 6, and 7	
PRI	4, 5, 6, and 7	
BRI (1B+D or 2B+D)	6 and 7	6 requires an MUX
POTS	6 and 7	6 requires an MUX

T1/E1 services. Also, the table does not reflect differences in the QoS such as AAL5 and AAL1 services that can have a large impact on the service offering.

Customer Location Qualification

The decision to install a host terminal into a building cannot be made solely on the basis of a single sale. Specifically, the single sale could be for a 256K Frame Relay circuit that may require more than $10K for implementation, installation, and capital equipment costs. The revenue from a 256K Frame Relay circuit over the life of the equipment cannot be justified unless there is another compelling reason to install the host terminal (for example, future sales).

The determination of future sales potential is effectively a more granular form of market and location estimation for bandwidth use. The structure being studied is in itself a mini-system that the operator uses to establish guidelines on whether or not to install the host terminal into the location.

The process is a multi-step approach that involves the following:

- Customer contact
- Completing a customer survey
- Evaluating co-tenants for potential bandwidth utilization
- Site survey (determining the entrance facility, service facility locations, and host terminal location)
- Estimating installation costs
- Determine current and future host terminal CI requirements

The purpose of the evaluation process is meant to ensure that the location selected for the host terminal installation is economically viable. At the same time, detailed information regarding co-tenants probably will not be available, resulting in estimates of the potential bandwidth usage for the complex. Finally, the process needs to be delegated to the lowest level possible in the organization, with a check-and-balance system implemented to ensure that the decision time and process is not too cumbersome and provides the necessary checks to prevent the inappropriate use of capital and company resources.

This is easy to say, but hard to realize in real life because the desire is to sign up customers as quickly as possible, especially in the early stages of the company's life cycle. Some facility evaluation can be relieved, based on the use of direct and indirect sales channels and the compensation methods utilized rewarding the sales force for business generated without installing a new host terminal.

Customer Survey

After the customer has been contacted and expresses a desire to potentially utilize the services offered by the LMDS operator, a series of questions needs to be asked of the customer. The type and volume of questions will directly help the technical staff with meeting the customer's expectations through properly meeting their requirements.

The following is an example of a customer survey that can be used by the sales team during the initial customer contact. The contact sheet (or survey) is meant as the first-level information. Obviously, if the sale is residential IP, the survey is not relevant except for the case in which the location of the equipment is to be installed, which is in the site survey process. Also, it would be ideal for a small or medium-sized business that the person in charge of telecommunications and IT be the one used to populate the survey with. But as reality dictates, most SMEs utilize a checkered approach to who is steering the telecom and IT requirements. Hence, it is important for the sales person to weed through the maze in advance.

It can not be overstressed that the sales team needs to be experienced with outside plant sales. The sales department also needs to have regular co-training sessions with the technical community.

This customer survey's purpose is to help formulate and guide your decisions regarding what network configuration the client company or office needs. The survey is also the first step in designing a new voice and data network or adding onto an existing one. The survey does not define the exact components that are needed for the network, but will provide the necessary information to ensure that as many issues are taken into account in the initial design as possible, from which a more detailed engineering work product can be generated.

Whether your computer system is currently networked or not at this time, there are a multitude of issues and configurations that your network can take on. The choices and perturbations seem endless: deciding if you need a simple peer to peer network, client server, LAN, WAN, extranet, or intranet. Coupled with these decisions lies the primary basis of whether to use a 10 baseT or 100BaseT with an ISA or PCI NIC; which will be connected to a hub, bridge and or router. Then, there is the protocol to utilize: will it be TCP/IP, IPX, NETBEUI, or something else? The number of choices and decisions that need to be made up front in the initial system design is daunting.

Of course, the primary issue that needs to be addressed first is what is needed in terms of a network configuration. The network configuration that is chosen will depend upon what applications need to be handled, coupled with the existing and future hardware that will exist in the company.

The configuration of the office network will need to take into account both the existing and future hardware and software that you or your company envision for the next one to two years. It is essential that the network configuration and topology is laid out right initially so that it can grow with your companies needs both from a application and financial viewpoint.

This document is meant to not only have you and your organization formulate what your current and future needs are but also to provide sufficient detail to allow a network design to be proposed that encompasses the necessary level of detail for proper cost estimates.

The first step in any process, whether it is going to the store or building a nuclear power plant, is to define what the customer wants to do, when it needs to be done, and what constraints are there being put in the process. Therefore it is essential at this stage to ask the primary question, what do you want to do?

Obviously, the question is meant to extract what customers' perceived needs are. For example, if they require everyone in the office to share the printer and thus eliminate the use of the printer switch and all the headaches that come with it, the use of LMDS-enabled technology would not meet their needs. Another example of what their needs are could involve order fulfillment. It may be necessary to have the shipping, order desks, and billing departments share information in real time, but they are located in separate campuses and require a WAN. Another example is if simple files and printers need to be shared at the same time to allow access to the Internet using a single Internet account. A fourth example is if executives and sales personnel need access from remote locations to the company's computer system for information, thereby setting up an extranet. There are many possible scenarios that can take place, but it is important to begin defining just what your client needs to have done.

In most of these examples, the LMDS system may be an enabler and the use of VARs or internal IT sales can be used as enhancers to the service offerings.

When discussing the client's requirements, it is of course important to bring up the advantages of LMDS service, which will meet their growth needs presently and also into the foreseeable future.

Table 7-2 is a brief form that can be used; the details can be expanded upon, based on the service offerings that will be made available. Although I have included many topics in this form, which may or may not be relevant to the LMDS system at hand, it is always easier to add and delete from an existing list than it is to create one from scratch.

NOTE: *The CIR and QoS associated with many of the data services will need to be reviewed as part of the package for determining the effectiveness of offering the LMDS service. It is important to note that price is not the sole determinant for a sale, and a combined package and bundling, will in most cases be extremely attractive to an SME and especially post sales.*

Physical Office Layout   An essential part of designing a network or adding onto an existing network is to account for the physical layout of the office and the service entrance information. The physical layout of the office needs to factor into issues such as the existing uses and requirements of the office. Are there multiple floors involved and does a physical separation exist between the various components that comprise the network?

The sketch should be crafted by the salesperson at the time of the client meeting to ensure that the information captured is relevant to the situation and requirements at hand. The sketch should be of the office layout as it exists right now, indicating the locations of the present computers, printers, telephone closet, and service entrance. Also, the drawing should show the approximate location of the electrical outlets within each room. Finally, the physical dimensions of each room need to be defined.

You should also indicate whether the ceiling is a drop ceiling for each room and the approximate height of the ceiling for each room. Other items to include involve identifying whether the computing area is on the same floor or whether different floors are involved. In the likelihood that an existing wire plant is present, you must include the location of the cross connect locations, hubs, or telephone closet where they all meet will be essential in the initial design.

Floor Layout Sketch

Table 7-2

Client contact form

Client Contact Form

XXX Sales Rep: Date:___/___/___

Client Contact Information:

Company Name_____

Address _____

City, State _____

Contact (name):_____

Phone:_____

Fax: _____

Email: _____

SIC Code: _____

What products and or services are offered at this location (business description)?

What specific applications, network functions, data, voice are needed now?

Do you have more than one location? If so, how many of these need to be considered in the network design?

The following information is needed to help quantify current and future requirements:

Voice Services	Current	Future (1–2 years)	Comments
Number of Employees			
Number of Extensions			
ISDN (BRI)			
Local Provider			
LD Provider			
International Provider			
Voice Mail			
Centrex (Y/N)			
PBX (Y/N) Type:			
Calling Features			
TR303/V5.2			
Trunk Types and Size (PRI,T1/E1 etc)			
Trunk Destination			

(continues)

Table 7-2

Client contact form
(continued)

Data/Network Services	Current	Future (1–2 years)	Comments
Number of Computers			
Number of Computers with Internet			
Number of Portable Computers			
Number of Fixed Computers			
Internet Access Provider		LMDS Operator	
Intranet			
Extranet			
ISDN (1B+D)			
ISDN (2B+D)			
XDSL (bandwidth)			
T1/E1 (where to)			
Fractional T1/E1 (n*64)			
Frame Relay (n*64) PVC			
Frame Relay (n*64) SVC			
ATM (Mbps)			
IP/Fax: Gateway Provider			
IP/Voice: Gateway Provider			
LAN			
Hubs			
LAN Switches			
Bridges			
Server			
Web Hosting			
Email			
E-Commerce			
RAS			
Operating System			
(NT, UNIX, Novell)			
Public IP Addresses (Range)			

Obtain Current Voice/Data Telecom Bills for Last Three Months

What do you want the network to do within the next two years?

How many computer stations, both location and remote, do you have now and will have in
the future?

Table 7-2

Client contact form
(continued)

	Present	6 Months	1 year	2 years
Local Computers				
Remote Computers				

When do you need to have the network or addition to the existing network completed by?
What are your timeframes?

What other items should be considered in the network design?

Future Bandwidth Estimation

Estimating the bandwith potential for a location is anything but easy. However, the following are some of the pieces of information that should be collected to establish the bandwidth requirements. The bandwidth requirements are largely driven by the types of companies that exist in the structure, how many people are at the facility (a parking lot is a direct indication), and the number of companies that are there.

When looking around the building to estimate the bandwidth, it will be important to note whether there is another LMDS operator at the location, whether any of the customers at the location are existing customers (a branch office), and the type of telco facility coming into the structure (copper, fiber, etc.). This usually can be extracted either from the potential client or a visit to the facilities or service entrance location.

The way to estimate the traffic for a complex is based on the assumptions used for the traffic engineering from the marketing plan. One might be compelled to provide this material to the sales force, but this would be ill-advised because of the lack of documentation control and the high likelihood that the information would fall into the competition's hands.

Therefore, Tables 7-3 and 7-4 are meant to steer the sales force to provide detailed enough data to the technical community. A brief capital authorization group, usually the sales manager, the person in charge of implementation, and the network engineering manager signoff on the building's approval.

The go-ahead should be based upon the need to deliver a particular Mbps of throughput to the location within a period of time, for example, one year.

Table 7-3

Type of structure

Item	Answer
Type	Office, Retail, Apartment, Warehouse
Floor(s)	
Fiber	(Y/N/Unknown)
LMDS Operator	(Y/N) Who:
Access to Service Entrance	(Y/N/Unknown)
Number of Colocated Companies	
Floor Space	
Vacancies	

Table 7-4

Client profile

Potential Client Profiles (for each company in building complex)	
Company	
Name	
Type (SIC)	
Floor	
Square Footage (m²)	
Number of Employees	(1–5)(6–10)(11–15),(16–20),(21–50),(+50)

The arrival of the Mbps can come directly from the marketing plan that identified a particular Mbps per customer type, etc.

For the sales aspect, the determination for building acceptance could use the quick evaluation shown in Table 7-5.

NOTE: *Score = Number × Weighting*

Table 7-5

Building score card

Type	Number	Weighting	Score
Offices		4	
Retail		2	
Service		3	
Apartment		1	
Industrial		1	
Total			>9 build

The actual weighting and scoring functions shown are for representation purposes, however the goal for LMDS is to have the host terminal provide services to as many customers as possible, or possibly one very large client. Therefore, the building's attractiveness for installation is directly dependent upon the amount of potential customers, which equates to bandwidth potential that exists at that complex.

Installation Cost Estimates

The installation estimates are rather straightforward, but need to be included for completeness. More specifically, the data required for approving the go-ahead regarding installation is to compare the estimated construction costs with those budgeted as shown in Tables 7-6 and 7-7.

The previous list can be considered a partial list and needs to be tailored to your specific requirements. However, if the cost to install exceeds the potential revenue from the facility over the expected lifetime of the product, then the economic viability of the installation needs to be reconsidered. The same argument can go for the monthly recurring costs. Hopefully, they will not exist but experience has shown that this topic cannot be avoided. For instance, if the monthly recurring cost is $150/month and the gross revenue from the service being delivered is $150/month, the economic viability needs to be reviewed.

Installation costs can also be reduced by the use of VARs, but this comes at the expense of losing direct control of the installation quality and the

Table 7-6

Capital costs

Item	Cost
Host Terminal ODU	
ODU Mounting Bracket	
Cable and Connectors, Bus Bar	
Lightning Arrestors	
IDU Host Terminal	
UPS and Cabinet	
Ancillary Platforms	
Installation Labor	
Engineering, Provisioning Optimization, and Testing Labor	
Total	$\leq$ budget

Table 7-7

Monthly recurring costs

Item	Cost
Lease Costs (if applicable)	
Power (if required)	
Total	$\leq$ budget

interface to the customer. Depending on the VAR chosen, the symbiotic relationship may prove extremely beneficial by leveraging the particular skill sets that best match the requirements at hand.

Customer Interface (CI) Requirements

The CI requirements from the host terminal need to factor in not only the current client under consideration, but also all the other potential clients that are in the same complex. Include the total service estimated for all the companies and then factor into a take rate, 25%, 50%, or 100%, for the building occupants to utilize your service.

From the take rate and the types of companies that exist in the complex, estimate the CI requirements plus the Mbps throughput.

For instance, if the host terminal that is used by the LMDS operator is capable of supporting IP only, the key issue is how many IP ports are available for connecting customers to it. At the same time, if the customer requires that voice traffic be transported, then VoIP may be the solution, provided the QoS can be secured within the LMDS system itself. VoIP can just be a method of providing voice services over an IP platform and then connecting to the PTT as a TDM circuit.

Future

The amount of permutations that are possible with the vast array of host terminals is perplexing and does not allow for any cost savings. The cost of deploying and operating an LMDS system is largely dependent upon the host terminal cost and flexibility. Some of the fundamental problems with the customer interface and host terminals are that all too often some of the services that need to be delivered to the customer do not match the direct capability of the host terminal. Thus, additional adjunct platforms need to be added to the host terminal in order to enable the particular service to be offered.

In order to reduce costs for the host terminal and to really be able to deliver broadband services to the customer, whether it is residential or commercial in nature, the host terminal needs to be standardized to allow multiple technology platforms to interface to it. The idea of an open platform for integrating lower-cost host terminals to the radio transport system would have a strong appeal for the LMDS operator.

The commonality could be achieved through agreed-upon personality modules that are interchangeable from vendor to vendor link, such as how a PCMCIA card can be exchanged computers. Figure 7-8 represents a possible multi-service platform for broadband LMDS.

References

Gramps, Peter. "Front-Ending Target Customers." *America's NETWORK*, February 1, 1995, 60.

Lewis, John. "Hubs and Networking." *Cabling Magazine*, May 1998, 20.

Mullen, Jim and Nancy Brandon. "Networking Networks: Wired or Wireless?" *WB&T*, December 1996.

Russell, Tom. "Cabling the LAN with T-1 to T-3 Circuits." *Cabling Business*, January 1996, 36.

Schindler, Esther. "Playing in the Vendor's Sandbox." Sm@rt Reseller, October 25, 1999, 44.

Schreyer, Andreas and Dave Schneider. "The Road to Safe and Effective VPN Solutions." *Telecommunications*, May 1999, 62.

Vyvey, Patrick. "MHz or MBPS." *Cabling Business*, January 1996, 18.

Implementation
Issues

Introduction

The implementation of an LMDS system can be relatively simplistic or extremely complicated, depending on the market, infrastructure utilized, and business plan. The majority of the LMDS systems have similar installation issues with regards to the base stations such as cellular, GSM and PCS. However, the host terminal installations are quite unique to LMDS and require a unique blend of installing the unit to last 15 years cheaply.

For LMDS systems, the most tricky part of the program is installing the required base stations in a just-in-time fashion that is at or under budget and will provide the needed coverage for maximizing the Mbps capability for the zone it is installed in. Although the ideal situation is what is desired, tradeoffs will always have to be made as real-life situations present themselves. Some of the obvious real-life situations involve site availability, ordinance approvals, equipment availability, and subcontractor coordination and control.

Implementation for LDMS systems involves traditional land use acquisition, commonly referred to as real estate or site acquisition. Implementation also involves the physical construction of the facility for use by the LMDS operator, as well as by clients at the host terminal location. The third aspect for implementation involves the provisioning and commissioning of the equipment at the base station, host terminal, or central office. Implementation can also include system expansion in terms of radio or sector adds to the base station or installing packs into an ATM switch.

This chapter will cover many of the salient issues associated with the implementation of an LMDS system. There will be variations based on the technology platform chosen, in addition to the host terminal and the services offered. However, many of the issues are the same, regardless of the configuration that is implemented for an LMDS system.

Land Use Acquisition (Real Estate)

Land use acquisition, commonly called real estate in the wireless industry, plays a critical role in wireless system designs and factors heavily in the design of an LMDS system. All too often, the design and future options that a network provider has at its disposal is driven not by technological issues but by issues related to real estate (land use acquisition). It is strongly suggested that because this area has such a strong and profound impact upon the network's design (present and future), the key issues discussed here should be known.

Site acquisition is the process that is utilized by the operator or potential operator to obtain base station, host terminal location, switching concentration center, office, or any issue related to real estate. Site acquisition involves the identification of the prospective property that meets specified conditions for potential use. Site acquisition can involve obtaining a lease, obtaining a right of way, or actually purchasing property.

The site acquisition process usually begins with the desire to establish a presence at a location. The desire to establish a presence can come from a variety of departments within the LMDS organization. However, the predominant group that utilizes the site acquisition department is the technical community.

The process begins with the generation of a search ring that defines the particular requirements. The process then enters into a search mode, where candidate locations are sought and compared against the engineering requirements.

One important aspect of LMDS is that unlike other wireless services, height is an advantage, within reason. Also, the site acquisition process may be heavily constrained, based on interconnect requirements or partnerships established. Once the site has been secured, the process quickly moves into the construction phase, which still involves the services of the site acquisition group.

However, once the base station is completed, the site acquisition group's involvement has not entirely ended, depending on the group's functions. Typically, the site acquisition group is responsible for ensuring that the lease payments and other particulars of the lease are enforced from both the LMDS operator's role as a lessee and as a lessor.

The steps and processes for finding an LMDS base station are different from market to market and also from company to company. However, there are many commonalities between the markets and the companies that can be exploited. The following is a general sequence of steps needed to be taken by primarily the real estate and engineering departments within an LMDS company for the purpose of securing a base station. Although the process may differ slightly for your company, the steps listed should apply to 90% of the issues that need to be addressed for an LMDS system when acquiring a base station.

The issues associated with implementing the RF portion of the LMDS system are listed as follows. The time duration that accompanies each of the steps is not included because they depend directly upon the size of the system as well as the time-to-market requirements. The following plan can be used for a new system and an existing system that is trying to expand by introducing a new technology platform to the network.

The process shown assumes that the RF design has already taken place, and the process begins at the issuance of the search ring itself.

Cell Site Acquisition Process

- Search areas issued
- Review search area requirements
- Review database for existing cell sites and friendly sites
- Obtain three candidate sites per search ring (if possible)
- Real estate contacts landlords
- Real estate coordinates site visits
- Physical visits (real estate/RF)
- Preliminary accept/reject
- Real estate reviews ordinances
- RF Engineering generates RF site visit information
- Real Estate generates site visit information
- Preliminary RF accept/reject for property
- Real estate starts preliminary lease negotiation
- Real estate/construction awards AE
- Real estate arranges construction and AE site visit
- Real estate/RF engineering/AE/construction site visit
- Real estate obtains lease exhibits from engineering
- Real estate obtains 60% drawing from AE
- Review 60% drawings
- RF Engineering performs FAA evaluation
- Review and accept lease exhibits
- Real estate receives preliminary lease
- Real estate obtains 90% drawings from AE
- Real estate reviews lease 90% with landlord
- Real estate submits lease to LMDS operator management
- LMDS operator approves lease
- Prepare building permit
- Request permit application from building department
- Obtain permits

Zoning (If Failure to Get Permit)

- Real estate awards attorney
- Real estate obtains variances required

- Real estate gets zoning application signed by landlord
- Real estate gets on planning board agenda
- Real estate assist attorney/AE/RF with prezoning
- Real estate/AE/attorney/RF attend work session
- Real estate instructs AE to revise plans and or application
- Real estate/AE/attorney/RF attend prep meeting
- Real estate/RF Engineering obtains EMF report

(Intervals estimated will differ depending on the board's agenda and the local ordinance requirements)

- **Wait one week**

  Real estate/AE/attorney/RF attend planning board meeting

- **Wait one month**

  Real estate/AE/attorney/RF attend architect review board meeting

  Real estate/AE/attorney/RF attend zoning board meeting

- **Wait one month**

  Real estate /AE/attorney/RF attend planning board meeting

  Real estate reports planning board accept/reject decision

  Wait for planning board resolution

- **45-day appeal process**

  Real estate reports results

  Real estate instructs AE to match resolution

  Attach resolution to building permit application

  Submit modifications to building dept

  Obtain permit

Search Process

The search process begins for the site acquisitions aspect with the issuance of a search ring. A search ring, also referred to as a candidate ring, is a document from the RF Engineering department, stipulating what the requirements are for the potential cell site that they want to establish for the network. A search ring can also be applicable to the MSC location or even a hub, to mention a few possibilities. However, a search ring is commonly associated with the cell site build program. Some of the materials commonly referenced in a search ring are the following items. Of course, the particulars for the potential location can change, based on the particular vendor's equipment requirements.

Search ring area

Date required to secure

Space required

AMSL and potential AGL

Number of antennas intended to be installed

After the search ring has been issued, the site acquisition team attempts to find site candidates. Site candidates are potential LMDS site locations that have been identified by either an internal or external real estate department for the company that meets the search area requirements specified in the search ring. The candidate selection is a process that involves the identification and subsequent approval of a potential site that meets the real estate, engineering, and construction requirements. Typically, there will be two or three site candidates identified per search ring, depending upon the requirements or constraints placed upon the search. The reason for more than one site candidate per search ring is because of the multitude of potential issues. In different stages of the sites negotiation or permit phase, Candidate A will no longer be acceptable while Candidate B might now be the best alternative to meet the design criteria. Often, the candidate selection will involve deciding which of three properties best meets the requirements from a leasing, RF coverage, and constructibility point of view.

Site Survey Form

A site survey form is one that is used by the initial investigation to the site to determine its validity. There are many variations, but Table 8-1 shows a simple form that can be used to convey the technical information back to the engineering community.

Lease Acquisition

A lease is one of the most-used vehicles for obtaining LMDS base station facilities in the industry. A lease is a written or oral contact between a land-

Table 8-1

Site survey form

Site Survey Form

Site Number_____   Date: / /XX

Address: _____

_____

_____

Obstructions (direction and type)

Terrain

Colocation (who/where)

Equipment Location

Land Use Issues:

Comments:

Roof/Antenna Sketch:

lord and a tenant, which transfers the right to use the property for a specific length of time in exchange for rent. The leases that are typically used for wireless, and in particular LMDS, involve written documents that also contain lease exhibits specifying the particulars of the site's use (in particular, the lease hold improvements that do take place).

The typical lease length is five years with three renewable options, which are at the discretion of the wireless operator. The initial five, plus three five-year extensions, give the lease a potential life of 20 years. Very few leases are signed for fewer than five years due to the extensive investment needed to improve the facility. There are exceptions to this, but they are very specific in that the facility is not envisioned to be usable in the near future and will be replaced with a more permanent location within the specified period of time.

Lease acquisition, also referred to as the leasing process, involves the activity performed by real estate for the sole purpose of obtaining the rights to use a particular property for wireless facilities. Some of the steps involved are as follows:

- Negotiation with landlord
- Lease exhibits
- Title search conducted

- Lease memo (if required)
- Local jurisdiction information defined
- Environmental studies (if required)
- Coordination with legal and provide input for lease language
- Obtain signed and executed lease

Types of Leases

There are numerous types of leases, and each type has specific issues. The types of leases that will be mentioned here are as follows:

- Sub-lease
- Tower lease
- Roof top lease
- Interior lease
- Exterior lease

Sub-lease

A sub-lease is a lease in which the lessor is leasing the location or part of the location for use by the wireless operator. It is important to identify because many roof top management firms are the master lease holders and offer leases to tenants—meaning that they are sub-leasing the facility from them. The importance is that the sub-lease must live by the agreement to the master lease; if the master lease is to expire in three years, signing a five-year lease with several renewable options may not be relevant. It is, therefore, worth checking to see what the master lease gives your potential landlord as rights and the terms for the lease.

Tower Lease

A tower lease is a lease that is associated with the location of the wireless communication system on a tower. The key issues that need to be factored into the lease involve the location of the antennas, the number of permitted antennas, maintenance, and accessibility to not only the tower but also the equipment room or shelter associated with the radio equipment. Another key issue that needs to be addressed is the location of the radio equipment, including power and security associated with access to the equipment. Then, there are compliance issues as well as procedures that need to be followed when another potential tenant wishes to locate on the premise due to interference analysis reasons and antenna placement.

Roof Top Lease

A roof top lease is associated with locating either radio equipment or antennas or both upon the roof of a building. A roof top lease may be a sublease if dealing with a roof top management firm. Some of the issues associated with a roof top lease that should be addressed up front include accessibility to the equipment and antennas for 24-hour availability and the location of the equipment, including power hookup. The antenna placement needs to be secured in advance to ensure that the optimal position that is available is obtained. Another issue is the procedure that should be followed when another potential tenant wishes to locate on the premises for the purpose of interference analysis and also to ensure that the other tenant's antenna system does not block or hinder the proposed or existing antenna structure secured in the lease. More times than not in the wireless industry, a roof top lease is part of the interior lease agreement that is worked out.

Interior Lease

An interior lease pertains to a *tenant improvement* (TI), and involves the radio equipment occupying space within a building, that is, interior. The lease also addresses the placement of the antennas and feedline runs, plus power and HVAC requirements. An important issue associated with interior leases is the availability of parking for the technician's vehicle, plus elevator access if the room is not on the ground floor. The interior lease is more common in urban environments.

Exterior Lease

An exterior lease pertains to raw land or to the placement of a shelter on a property. The exterior lease defined here is one in which the equipment is located outside, and has its own cabinet or shelter requiring no modifications to the existing building, with the exception of antenna-installation issues. Some of the more common issues of an exterior lease pertain to access and parking for operations personnel as well as the issue of antenna placement. One important item for exterior installations pertains to the availability of a generator hookup.

Gross Lease

A gross lease is one in which the landlord pays all the property expenses normally incurred, such as taxes and regular maintenance of the building or property. For wireless, the gross lease is a common type of lease.

Ground Lease

A ground lease involves a lease for the ground only. The tenant owns the building or constructs a building over the existing land. The ground lease is a typical for green field applications.

Net Lease

A net lease stipulates that the tenant pays not only the rent but also some, if not all, the operating costs associated with a property. Some of the additional operating costs for the property could be taxes, the utilities for the building, and the repairs required. The net lease is important for wireless operators to be on guard for because this is where the full loaded cost of the property can be passed to the LMDS operator.

Common Lease Terms

There are numerous issues that arise out of even the most basic lease. This section addresses many of the more salient points regarding common lease terms that should be known by upper management when reviewing the lease for approval.

The designation of who is the lessee and who is the lessor is often confusing at first. Lessee is the term used to describe the tenant in a lease. An example of a lessee in wireless is the LMDS operators themselves. A lessor is the landlord in a lease.

Rental rate is exactly what it implies—the agreed-upon amount of money or similar compensation that is owed the landlord (lessor) by the tenant (lessee) at the predetermined interval that can be monthly, yearly, or whatever is agreed upon in advance.

COLA

Cost of Living Adjustment (COLA) is a common term that is inserted in a lease to indicate the amount of rent increases expected to be incurred by the wireless operator. COLA is either a fixed amount, say 5% per year, or it is based on the *Consumer Price Index* (CPI). The value and methodology used for COLA should be examined in detail because a 5% increase over the course of the life for the site may seriously increase the operating site expense.

Tenant Improvement

A tenant improvement is associated with the process of improving the area within an existing building for the purpose of housing cell site equipment. This is usually the interior fit up process and the result of preparing the equipment room that is leased to house the radio equipment. Some issues associated with tenant improvement involve the separate power feed, HVAC system, and antenna cable runs.

Access

Simply put, access is the ability to enter and leave a facility at will. Usually, access for a wireless facility is defined as unlimited in that there is 24-hour access, seven days a week, all year round. The reason access is so important for wireless facilities is to allow for potential maintenance to take place, both preventive and emergency issues, which could be the replacement of a defective radio or antenna system. Access rights might however be slightly different for roof tops because they are for interior locations. Usually, the roof access is not as critical because this is related to the antenna system, and repairs to them are less frequent. However, if the equipment is located on the roof in a cabinet enclosure, then the access to the roof becomes paramount to be unrestricted.

Sub-meter

A sub-meter is separate electric service that provides electricity to the cell site. The advantage of the sub-meter is that the actual usage used by the cell is billed directly to the wireless operator. Care must be exercised to ensure that the monthly bills actually make it to the operator and do not sit in the corner of a room waiting for the utility company to disconnect service through non-payment.

Check-Meter

A check meter refers to the situation in which the operator or landlord checks the usage that is on the electric meter associated with the cell site. It is important to note that the check meter is not associated with the utility company as such, and is directly associated with the landlord. It involves a visual checking of the meter and then a separate payment process worked out with the landlord.

Option

An option is simply an agreement to keep open the ability to either lease or purchase the property in question for a defined period of time. Options are commonly used in wireless as a method of securing properties in advance of their actual use. The use of options are meant as a method to secure the property while the land use acquisition or zoning process takes place to obtain the necessary building permit. The options, however, typically have an expiration or termination clause, and these clauses and expirations should be noted from the beginning. In several cases, the zoning process was well underway when the option was expired, causing the potential facility to be abandoned after much time and effort, not to mention money, was expended.

Title Search

A title search is a process by which the public records are searched with regard to the property in question to determine the ownership and any encumbrances affecting the property. It is strongly advised that before any wireless operator enters into a five-year lease, or any lease with renewable options, a title search be conducted for the property. The cost of the title search is trivial when compared to the economic investment that will take place with the property.

Survey

A survey is the process by which a parcel of land is measured and its area ascertained and shown on a map, indicating the boundary measurements. Often, a survey is accomplished when constructing a shelter or installing a cabinet on the property. A survey is also required for a building permit. In many cases, however, the level of detail required for the survey is dependent upon the local building inspector's requirements.

Assignment

An assignment is commonly associated with a lease, but it could be applied to a bond or mortgage to mention a few. An assignment is the case in which the interest in a given property can be transferred from party A to party B. This is important when the building is sold and you lease it; you may or may not want the lease to be assigned to the potential new owner. The same issue goes for wireless operators who appear to be caught up in the merger and acquisition frenzy, in which the lease issues are assigned to the new owner.

When the transfer of ownership occurs, it is important to avoid renegotiating the lease and potentially either losing the property for its current use or having an increase in the lease rates.

Month to Month

The month to month lease is really a month to month tenancy and references a time frame for one rent period at a time. This is usually seen when there is no formal agreement between the tenant and the landlord for use of the property or during a transition period. The month to month tenancy does not have to be month to month; it can be for a three-month interval, although the common period is monthly.

Title

In real estate, a title is evidence that the owner of land is the actual owner. Why this is important to wireless is that the title will determine whether the person you are negotiating with is the actual owner or is an agent of the owner. It is possible for an unscrupulous individual to offer the property for use and attempt to obtain money when it is not under their control. Also, a title will indicate any outstanding liens that could also be placed against the structure. However, the ownership issue is the most critical first step.

Quiet Enjoyment

Quiet enjoyment is a term that is commonly in many leases and is the right of a person, either the owner or lessee, to use the property without interference of any kind. Where this is applicable to wireless is the issue of physical aesthetics, antennas, and noise caused by the HVAC system or fans from the cell site. The quiet enjoyment issue is not normally invoked by the wireless operator; it is usually invoked by other tenants who are colocated with the wireless facility.

Approvals

The permits needed for constructing a cell site vary from municipality to municipality. Many times, to obtain a permit, a variance to the local zoning codes needs to be obtained first if the construction and/or use is not granted by right. The place to determine the specific requirements for obtaining the building permit and ancillary permits that also may be required is under the direct purview of the building inspector. It is strongly suggested that in the initial feasibility phase of the cell site's life, the actual requirements for permits be determined to avoid any problems in the future.

The local ordinances for the municipality where the installation is being sought define what is required for a permit to be issued. The local ordinances are rules and regulations that define what is allowed to be constructed or rather define what uses are permitted in a given area for the municipality. The local ordinances, zoning rules, usually provide the guidance as to what can and cannot be built by right or a variance is required to operate there. If a variance is required, the variance request proceeds through the local boards for the necessary approvals.

The local ordinances are defined in the building codes for that municipality and are the regulations that are set forth by the state and local authorities to ensure that the structural and electrical requirements for a given structure are meet. Each town has their own set of building codes and these building codes often determine what is required when building a site. Some examples of building codes are the Americans with Disabilities issues, in which access to and from the site is defined. Another issue is whether the site is a manned or unmanned facility. If it is manned, toilet facilities may be required as compared to none if it is an unmanned facility.

There are numerous issues associated with zoning and land use approvals from wireless operators and real estate perspective. The issues faced are unique for each location and municipality encountered including the differences between building departments.

The following gives a general list of some of the issues that will need to be addressed during the zoning and land-use approval process.

- Establishment of a zoning/land use strategy
- Submitting various zoning and land use applications
- Obtaining the required zoning and land use approvals
- Coordinating environmental assessments (if required)
- Obtaining and altering site plans when required
- Attending and potentially testifying at planning and/or zoning board hearings
- Coordinating any third parties needed for the LMDS operator when required
- Obtaining the building permit for the base station

Building Permit

A building permit is a permit given by the local government, usually the building inspector, for the purpose of constructing the proposed site. In some cases, the building permit is secured (that is, it's authorized in the particular building code). In most cases, however, the building permit can be obtained only after the approval of the planning board. The particular issues and sequences associated with obtaining a building permit differ from municipality to municipality. In all cases, however, the requirements for obtaining a building permit are obtained from the building department of the local municipality. It is recommended that you obtain a building permit prior to the commencement of construction for the base station.

Variance

A variance is permission obtained from the local zoning authority to build a structure, or to use a structure or conduct a use that is expressly prohibited by the current zoning laws. The variance is basically an exception from the zoning ordinance. A variance is often required for wireless applications with regards to height or use because most municipalities do not grant the ability to operate a wireless facility by right. Also, the usual height restriction pertains to a two-story home or a 65-foot silo for a farm, and in almost all cases the wireless facility requires more height than that allowed. Please remember that the average tree line is 70 feet in height.

Zoning

Zoning refers to the ordinances stipulated by a municipality that directly influence how a wireless operator can possibly construct a wireless facility in that community. Although zoning takes on many potential meanings, the fundamental principle is that the local zoning ordinance will specify whether a cell site can be constructed in a given area or not. Usually, the ordinance will not specify a wireless facility; in this case, the normal reading is that if the use is not permitted it is by default excluded, requiring a variance to construct and operate the cell site. The ordinance stipulated in the zoning guidelines will reference several types of zones. They are usually industrial, commercial, or residential. There are, of course, other variations to the three zones listed, but all municipalities utilize the three general classifications.

Residential Zone   A residential zone is an area, defined in the local ordinance book for the particular municipality, which specifies the geographic boundary where the use for any building in that area is for residential use only. There are differences in the type of residence—they are usually referenced in terms of R1, R2, etc.—and each has a different stipulation and set of requirements. Usually, if a base station is selected by engineering to be located in a residential zone, the opposition from the local community will be great. It is usually preferred to locate the potential cell either on an existing tower or in a commercial or industrial zone for the community. This may be different due to the market the LMDS system will serve.

Commercial Zone   A commercial zone is an area, defined in the local ordinance book for the particular municipality, which specifies the geographic boundary where the use for any building in that area can be used for commercial use. The allowed commercial use is defined in the local ordinance requirements (an example of commercial use is a convenience store).

Industrial Zone   An industrial zone is an area, defined in the local ordinance book for the particular municipality, which specifies the geographic boundary where the use for any building in that area can be used for industrial use. The allowed industrial use is defined in the local ordinance requirements (an example of industrial use is a factory). However, there are degrees of industrial use. Often, when a community defines an industrial zone, it either allows heavy manufacturing (such as textile plants) or refers to a light industrial. Why the industrial zone is important for wireless operation in a particular LMDS is that the ability to construct a cell site is usually met with minimal resistance if the facility is able to be built in an industrial zone.

Landmarks   Landmarks is an area within a community defined for historic preservation, requiring unique aesthetic requirements for the construction of improvement to any building in that defined area. Landmarks also is a term used to describe the name of a Landmarks Preservation board or Landmarks board for the community.

Architectural Review Board

The architectural review board is one of the local boards from which approval may be needed in the process to obtain a building permit for a potential cell site. The architectural review board often oversees that type of structure and how it is designed to fit within the community. The specific requirements and authorization that the board has is different from municipality to municipality. However, often the placement of the antennas for a wireless facility will be reviewed and potential alterations will be made to satisfy the board's desire to obtain a favorable vote that will be sent to the planning board for the town.

Certificate of Occupancy

A *certificate of occupancy* (CO) is needed for every base station, switching facility, and any other facility used by the wireless operator. The certificate of occupancy is a piece of paper or document that is issued by the local municipality, stating that the building or facility complies with the local and state building codes, as well as the health and safety codes. The CO enables the facility to be occupied; without it, you are not allowed to occupy the location, even if you are just performing maintenance. This is a critical document and is often overlooked by operators because it is obtained after the site is operational.

Colocation

Colocation refers to the situation in which an LMDS operator will occupy the same facility, building, or tower as another wireless or CLEC operator. Colocation is becoming more common in the industry to facilitate the operator's build program of adding cell sites to the network within a defined period of time. Colocation, however, is extremely prevalent in the fixed network component of the LMDS system with the stated purpose of reducing interconnect cost by locating the fixed network equipment at the point of presence of a CLEC.

A colocation procedure that has been used for coordination between other wireless companies is briefly defined as follows and can be expanded upon. The durations, of course, can be modified to fit the situation.

1. Real Estate

   ■ Initial contact when a colocation situation is determined (verbal)
   ■ Real estate contacts exchanged
   ■ Three business days can elapse before escalation procedure begins

2. Engineering—Review/Approval

   ■ Engineering contacts begin with a verbal description of what is desired
   ■ Site particular parameters exchanged between both service providers
   ■ Sample lease exhibit drawing generated and submitted to existing operator for review and comment (this is in memo form and is faxed) (5 days for review)
   ■ Operator initials drawing for approval
   ■ Six business days can elapse before escalation procedure begins
   ■ Generate revised lease exhibit drawing and resubmit to operators for their records

3. Construction—

- Two weeks prior to commencement of construction, contact existing operators used for engineering review and notify them of impending action (real estate contact)

Installation (Construction)

The rapid growth of the broadband wireless industry is exhibited, not only in terms of the amount of subscribers that now utilize wireless, but also by the sheer number of wireless sites that are currently in operation or will be in the near future. The construction aspects for an LMDS system are critical and in many instances have a direct impact on the ultimate performance of the system.

Construction is the phase that represents the group or action that takes place for constructing a cell site. The construction phase is part of the implementation process. However, some operators refer to the construction as the implementation process. Construction is the actual modification of a location so that the wireless equipment can be installed and operated properly at the defined location. Some simple examples of construction involve the tenant improvement process or the erection of a tower with the associated shelter.

When discussing wireless systems, installation and construction are usually centered on the base stations and the switching complexes. However, LMDS also has the issue of installation for the host terminals in a variety of building configurations.

Configuration

There are numerous configurations that are utilized for both LMDS base stations, host terminal installations, and switching complexes. The configurations are dependent largely upon the technology transport mechanism that is chosen by the LDMS operator. However, as the amount of variants grow, the more common elements increase.

The three major platforms for construction in an LMDS system are as follows:

Base station

Host terminal

Switching (packet) complex

Base Station Configurations

As suspected, there are numerous types of base station configurations ranging from roof top to tower or monopole installations. The physical installation can involve a single sector or multiple sectors each. Also, each sector can have different radio populations, which either require a separate ODU or a single integrated ODU in which multiplexing of the RF energy occurs either at the RF or IF level.

A typical LMDS base station or communication site consists of the components shown in Figure 8-1.

The equipment room for the base station shown in the figure can be in a building, located in a shelter on a roof of a building, or in an equipment shelter next to a communications tower. The HVAC system, cable trays, fire-suppression system, alarm system, cable entrance, lightning protection, and

Figure 8-1
Typical LMDS
equipment room

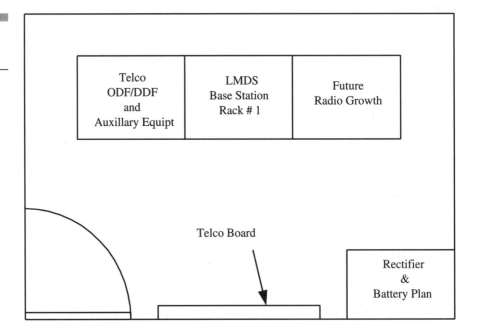

grounding are not shown in the diagram, but will need to be included with any proper design.

The corresponding outdoor equipment that accompanies the indoor equipment is shown in Figure 8-2 for quick reference. The figure is meant only as a generic diagram because there are a plethora of implementation issues associated with any installation. However, the typical LMDS configuration is the 4-sector site, 90-degree sectors.

Monopole

A monopole is a type of tower structure that is used extensively by cellular/GSM and PCS wireless communication for the installation of the antenna array utilized. The monopole is a single pole, hence the name, which has a top hat on the top used for mounting the antennas to it. Although the monopole's top hat can come in a variety of sizes, depending on the design requirements, a typical top hat is either 10 or 15 feet wide on any of its three typical faces. A monopole can also have several top hats associated with it to facilitate multiple colocation possibilities; hence, multiple carriers can easily use it. See Figure 8-3.

Self-supporting Tower

A self-supporting tower is one that requires no additional support, such as a building or guy wires. The self-supporting tower is usually the most expensive type of tower, but has probably the most flexibility in terms of

Figure 8-2
Typical LMDS 4-sector antenna site

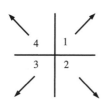

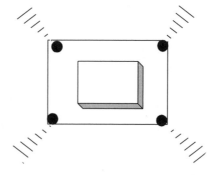

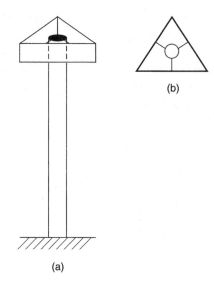

Figure 8-3
Monopole
a) side view
b) top view

(b)

(a)

configuration possibilities. An example of a self-supporting tower is shown in Figure 8-4.

Guy Wire Tower

A guy wire tower is a tower that requires a series of cables to keep it erect and help distribute the structural load. The guy wire tower is usually the cheapest to construct, but requires more physical space due to the placement of the support wires. Figure 8-5 shows an example of a guy wire tower.

Host Terminal Installation

The host terminal installation issues are unique to personnel coming into the LMDS arena from a mobility background. Specifically, the host terminal installation is like a microcell that needs to deliver service to multiple locations within a building or complex from a single radio feed point. However, the key difference is that this service is meant to interface to the fixed landline portions of the system, not supply mobility, either directly or through a wireless PBX alternative.

The host terminal installation not only has to address LOS issues, but also where the entrance and service facilities are located with respect to the client(s) for the complex being considered.

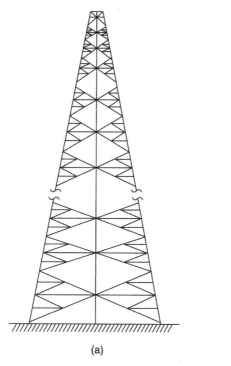

Figure 8-4
Self-supporting tower
a) side view
b) top view

(a)   (b)

Figure 8-6 attempts to depict the multidimensional issue of where the existing PTT/CLEC entrance facility is in relationship to the service entrance that is currently used by the client. The entrance and service locations do not need to be the same location, and it is important to account for this issue.

In particular, at issue are the wires or media used to connect the service entrance from the demarcation point in the entrance facility. More specifically, you need to know who owns these wires or media. Can you utilize them or do you need to pull more feeds to accommodate the current as well as future requirements of the client?

Figure 8-7 is a depiction of the service entrance (sometimes referred to as the wiring closet) within the complex. The location of the host terminal is included in the drawing, and it assumes that there is adequate space to install the unit by itself with separate power source, earthing, etc. In reality, however, the various wiring closets are small, resulting in installation constraints as well as cooling limitations, which should never be overlooked.

Another sore point that always arises is the protected power issue. Protected power, from both a surge as well as tampering aspect, needs to be included with the design. Finally, the physical protection of the host terminal also needs to be factored into the design. A simple rule to follow is that

▬▬ ▬▬ ▬▬ ▬▬

Figure 8-5
Guy wire tower
a) side view
b) top view
c) compound

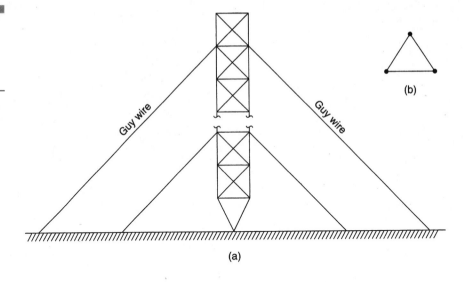

(b)

(a)

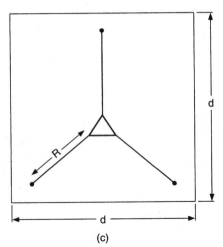

(c)

if it can go wrong, it will go wrong. Do not trust that the equipment will not be tampered with at some time.

Figure 8-8 shows a typical configuration that an LMDS operator may have to encounter when installing at a office building.

The figure assumes that there is only a single service entrance for the whole building, and that the service entrance is one and the same as the facilities entrance.

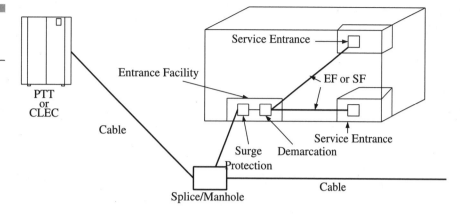

Figure 8-6
Entrance facility and
service entrance

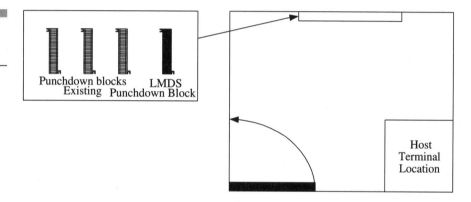

Figure 8-7
Service entrance
room (wiring closet)

Figure 8-9 is an LMDS installation involving row houses, or it can also represent the situation in a strip mall where all the buildings are on a single level. The installation depicted shows a single radio installation for the host terminal, which then feeds multiple clients. Each of the clients has its own entrance facility, thereby requiring external cabling to take place from the primary entrance facility to the secondary entrance facilities.

The illustration in Figure 8-10 shows the situation for LMDS installation when installing at a residential home. The installation shown in this diagram involves a single residence.

Finally, Figure 8-11 is a depiction of a multiple occupant building, namely a high-rise apartment building in which one host terminal provides the bandwidth needs for the entire building. A typical service offering could be high-speed, always on, Internet access.

Figure 8-8
MDU–Multiple
Dwelling Unit

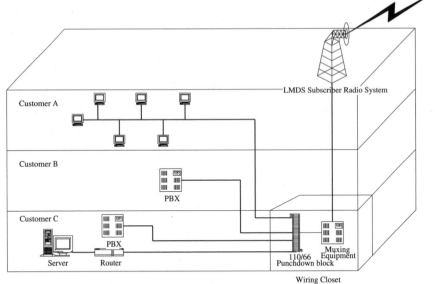

Multiple Customer Location/Dwelling Unit

Figure 8-9
Row buildings or
strip mall

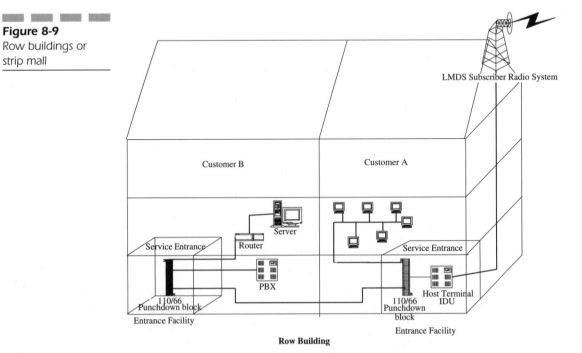

Row Building

Figure 8-10
Residential

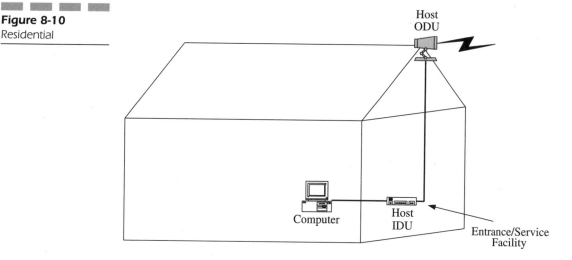

Figure 8-11
*High-rise apartment
building*

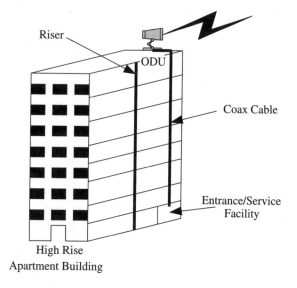

Switching Complex Layout

The layout for a switching complex is dependent upon what services and functions the facility is meant to provide. The layout will depend on normal variables such as the vendor(s), the ancillary equipment installed, and future expansion issues.

Figure 8-12
Packet-switching
complex

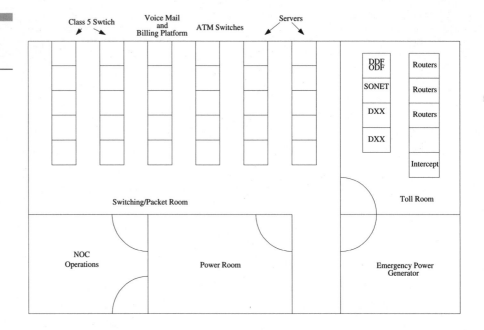

Figure 8-12
Packet-switching
complex

The determination for redundant power feeds, as well as interconnect facilities will also have a direct impact on the structure's configuration.

A typical MSC layout is shown in Figure 8-12 for reference. However there are numerous perturbations for MSC layouts, and they depend on the vendor used, the ancillary equipment installed, and future expansion for the location.

Equipment Structures

Equipment structures can take on many different views. The equipment structure, whether it is for an interior or exterior application, is meant to house the telecommunications equipment, usually radio equipment. Some of the common items that are normally found in a equipment structure or shelter involve the radio equipment itself, batteries, HVAC, rectifiers, and telecom demarcation equipment. The size and dimension of a shelter is quite varied because there are no defined dimensions. The goal is to keep the structure requirements as small as possible, yet allow for potential changes in the future such as technology improvements or vendor replacement. Equipment structures, shelters, come in two general varieties: interior and exterior.

An interior structure is another name for a *tenant improvement* (TI) site. The interior cell site occupies a particular amount of space within a building and has certain access privileges associated with it. The layout used by the LMDS operator should be standardized to prevent the design a la mode process. Of course, the standardization needs to have a certain inherent flexibility because each interior TI room has its own set of issues associated with it. An example of an interior install is shown in Figure 8-1.

The counter to the interior structure is an exterior structure. For the exterior structure, the use of a shelter, hut, or small self-contained cabinet is used. Obviously, the use of the term exterior site has many meanings, but it is not located within an existing building.

Antenna Structure

The antenna structure that is utilized at either the base station or host terminal location will directly influence the performance of the LMDS system. There are obviously other issues, but most of them are more easily adjusted or corrected as the system develops. The mounting of the antennas needs to be done with extreme care. Therefore, the following items should be checked off prior to acceptance of a cell site.

- Number and types of antennas to be installed
- Maximum cable run allowed
- Identify and rank obstructions that would alter the desired coverage (because this is an LOS system, obstructions are to be avoided)
- Fresnel Zone requirements are met
- Isolation requirements are met with other services
- Antenna AGL requirements are met
- Antenna-mounting parameters are met
- Intermodulation analysis is complete
- Path clearance analysis is verified

When installing on a tower, the physical spacing, offset from the tower, must be selected so that the tower's structure either enhances or does not

Table 8-2

Standard
conversions

From	To	Multiply By
Meters	Feet	3.28
Feet	Meters	0.3048
Miles	Kilometers	1.609
Kilometers	Miles	0.6214
Kilometers	Feet	3281
Feet	Kilometer	0.0003408
Liters	Gallons	0.2642
Gallons	Liters	3.785
Rods	Feet	0.06061
Yards	Feet	3
Yards	Meters	1.094
Inches	Centimeters	2.54
Centimeters	Inches	0.3937
Feet	Centimeters	30.48
Centimeters	Feet	0.03281

alter the antenna pattern desired. Also, whether installing antennas on a building or tower, the use of downtilting needs to be factored into the installation. More specifically, when stacking antennas (because LMDS service can be tailored by channel, not only by sector), a particular radio may need to be downtilted while the rest are not. This requires the necessary clearance being incorporated between antennas to ensure that it can be facilitated.

This list is just preliminary and can easily be altered based on the situation at hand. It should be modified to met your particular system design requirements.

When installing antennas on an existing roof or penthouse, you should take into account how high the antenna must be with respect to the roof surface. Obviously, the ideal location is to place the antenna right at the roof edge. However, placing the antenna at the roof edge may not be a viable installation design. When the antennas cannot be placed at the edge of the roof, a relationship between the distance from the roof edge and the antenna height exists.

The basic issues that need to be factored into determining how far back the antenna can be from the roof edge depend on the following:

1. Fresnel Zone clearance

2. Downtilt capability (without violating the Fresnel zone)

3. Setback requirements

Obstructions that can cause Fresnel violations do not only include the parapet wall, but also the HVAC system, window washing apparatus, existing or future antenna systems, and various chimneys or other such structures.

Figures 8-13 and 8-14 represent two roof top antenna installations. Figure 8-13 shows a vertical stacked system; Figure 8-14 shows a horizontal stack system.

Antenna Installation Tolerance

The antenna installation tolerances apply directly to the physical orientation and plumbness of the antenna installation itself. There are usually two separate requirements: how accurate should the antenna orientation be and how plumb should the antenna installation be. The obvious issue is not only the design requirements from engineering, but also the practical implementation of the antennas for cost reasons.

Table 8-3 shows the guideline that should be used.

The antenna orientation tolerance is a function of the antenna pattern and can be unique for each type of cell site. Obviously, for an omni cell site there are no orientation requirements because the site is meant to cover 360 degrees. However, for a sector or directional cell site, the orientation tolerance becomes a critical issue. The orientation tolerance should be specified from RF Engineering; in its absence, the guideline is to be within 5% of the antenna's horizontal pattern. Table 8-4 helps to illustrate the issue by using some of the more standard types of antenna patterns used in the industry.

The obvious goal is to have no error associated with the orientation of the antenna, but this is rather impractical.

Figure 8-13
Vertical stacked
antennas

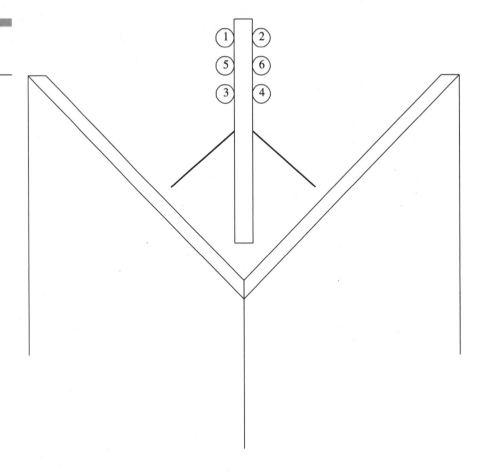

As the antenna pattern becomes tighter, the tolerance for the orientation error is reduced. The objective defined here is +/−5%, but the number can be either relaxed or tightened, depending on your particular system requirements. The 5% number should also factor into any potential building sway that does occur (usually a non-issue due to the height of the buildings used for wireless installations).

The vertical tolerance for an antenna installation involved with wireless communications is a tight and rigid requirement that is often poorly documented. The tolerance needs to be tight because of the direct impact on the coverage of the cell site. Too lax a vertical tolerance could have the same impact as the downtilting the cell site. The vertical antenna tolerance, plumbness, is +/−1 degree from true vertical.

Figure 8-14
Horizontal stacked
antennas

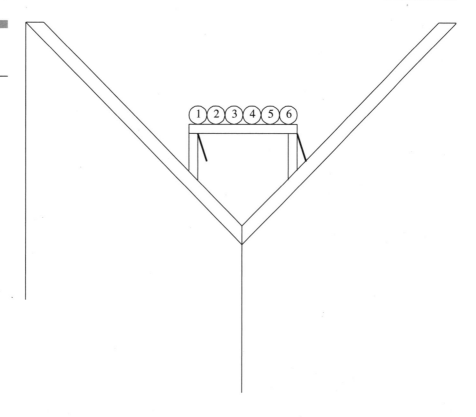

Table 8-3

Antenna
installation
tolerance

Type	Tolerance
Orientation	+/−5% of antenna's horizontal pattern
Plumbness	+/−1 degree (critical)

Table 8-4

Antenna
orientation
tolerance

Antenna Horizontal Pattern	Tolerance from Boresite
90 degrees	+/−4.5 degrees
45 degrees	+/−2.0 degrees
30 degrees	+/−1.5 degrees

Power

Power for a base station, switching complex, or host terminal needs to be designed and factored into the installation. In all cases, the use of DC power is required for normal telecommunications equipment (usually −48V DC), and it requires a rectification from AC power, either 110 or 220V volt (single or three phase). Coupled with the power requirement is the need for protecting the equipment power in the unlikely event of a power outage through battery backup systems. Along with power consumption, the byproduct is heat, which needs to be treated either by forced or convection cooling methods.

All too often, questions arise regarding power conversions due to the vast array of different measurement units utilized. Therefore, the following power conversions listed in Table 8-5 should prove to be helpful.

The rectifier plant must not only address the equipment requirements, but also the recharge rate for the batteries.

Batteries

Batteries are an integral part of a wireless communication system. There are batteries at the cell sites and at each switching complex within a wireless network. Often, ancillary equipment also has batteries associated with it. The batteries provide continuous electrical power to the system when there is an interruption in the commercial power system. There are numerous types of batteries, but the two most common types are the wet cell and sealed versions. The wet cell versions are more traditionally used at central

Table 8-5

Power conversion table

From	To	Multiply By
Horsepower	BTU per minute	42.418
Horsepower	Kilowatts	0.746
Kilowatts	Horsepower	1.341
Kilowatt hours	BTU	3413
BTU	Kilowatt hours	0.000293
Watts	BTU per hour	3.413
BTU per hour	Watts	0.293

office. They have an exceptional life, but require regular maintenance over and above keeping the batteries charged and the terminals clean. The sealed versions allow for more possible installation configurations, including many in-building applications. The sealed versions do not vent gases and therefore do not require the used of a hydrogen gas-venting system. The sealed batteries do not have the same shelf life as do the wet cell batteries, but require far less attention.

A battery string is a series of batteries connected together for purposes of supplying 48 volts DC to a communication system. Depending on the amount of amp hours required from the communication site, the number of battery strings will vary. For example, if it is determined that two strings of batteries are required for four hours of battery backup power, possibly going to eight hours of backup power would require an additional two battery strings, for a total of four hours of battery backup power.

HVAC

All LMDS telecommunication equipment has some level of *heating, ventilation, and air conditioning* (HVAC) requirements. The particular HVAC requirement is usually stipulated in terms of the BTUs required to remove, which is then equated to a particular tonnage for air conditioning. The primary use of the HVAC system is to remove heat from the equipment room. In some cases, however, heat may also be required because all the equipment, including batteries, is designed to operate within a given temperature range. Often, the HVAC requirements may require an upgrade to the power plant for the cell site or MSC.

Heat dissipation is the process or phenomenon of heat being dissipated, given up, by the equipment in a cell site or MSC in the course of being one and processing calls. The heat dissipation for each of the pieces of equipment used either at the cell site or the MSC is obtained either from physical measurements or provided from the manufacturers. The heat dissipation is usually defined in terms of BTUs, and this defines the HVAC requirements for the location.

A *British Thermal Unit* (BTU) is a representation of heat. The use of the BTU number applies directly to the dimensioning of the HVAC system for the cell site. whether it is an interior or exterior location. For cabinet installations, if the equipment comes preinstalled from the factory of the manufacturer (for example, Nortel, Motorola, Lucent, or Ericsson), the BTU dissipation requirements have already been factored into the design of the cabinet.

$$1\ \text{BTU}\ =\ 2.930\ \times\ 10^{-4}\ \text{kW} - hr.$$

LMDS Base Station Site Checklist

Some of the installation issues encountered at a site involve a multitude of issues. They are very numerous because each building or installation has some problem(s) associated with it. Simple issues that always are involved with installation deal with the coordination of the various contractors used for the installation process. In many instances, the dependence is serial—Project B cannot begin until Project A is completed, for example.

Following is a brief checklist of some of the more common installation issues that are encountered:

- Access to the site
- Hours allowed for construction (usually a tenant-improvement issue)
- LMDS base station equipment delivery
- Telco acceptance or installing a PtP radio link
- Securing the necessary permits
- AE drawing approvals from internal groups as well as the landlord
- Antenna system installation
- Landlord claims of damage to the building as a result of the installation work
- Power system upgrade
- Floor loading (primarily for batteries)
- HVAC venting and installation
- Noise abatement for the cell site and HVAC system
- Alternative power requirements (that is, generator hookup)
- Parking and bathroom facilities

To help prevent implementation problems of LMDS operational issues, the Table 8-6 contains a brief summary of the major items that need to be

Table 8-6	Topic	Received	Open
Installation checklist	**Site Location Issues:**		
	24-hour access		
	Parking		
	Direction to site		
	Keys issued		
	Entry/access restrictions		
	Elevator operation hours		
	Copy of lease		
	Copy of building permits		
	Lien releases obtained		
	Certificate of occupancy		
	Utilities		
	Separate meter installed		
	Auxiliary power (generator)		
	Rectifiers installed and balanced		
	Batteries installed		
	Batteries charged		
	Safety gear installed		
	Fan/venting supplied		
	Facilities		
	Copper, fiber, PtP		
	Power for fiber hookup (if applicable)		
	PtP radios aligned		
	POTS lines for operations		
	Number of facilities identified by engineering		
	Spans "shaked and baked"		

checked prior to or during the commissioning of a communication site. The checklist is generic and should be tailored for your particular application (you can add or remove parts where applicable). It is an excellent first step in ensuring that everything is accounted for prior to the communication site going commercial.

Table 8-6

Installation checklist
(continued)

Topic	Received	Open
HVAC		
Installation completed		
HVAC tested		
HVAC system accepted		
Anetnna System		
FAA requirements met		
Antennas mounted correctly		
Antenna azimuth checked		
Antenna plumbness checked		
Antenna inclination verified		
SWR check of antenna system		
SWR record given to Ops and Engineering		
Feedline connections sealed		
Feedline grounds completed		
Operations		
User alarms defined		
Engineering		
Site parameters defined		
Interference check completed		
Installation MOP generated		
FCC requirements document filled out		
Optimization complete		
Performance package completed		
Radio Infrastructure		
Bays installed		
Equipment installed according to plans		
Radio equipment ATP'd		
Tx output measured and correct		
Grounding complete		
Equipment bar-coded		

Grounding

Grounding is an area that inspires many debates and is a large source of error for communication site installations. Grounding specification quality varies dramatically from vendor to vendor as the level of engineering experience is applied to practical implementation issues, and not just a reference to ground the equipment.

The purpose of the grounding system is to not only provide personnel safety from the threat of electrocution, but also to protect the electronic equipment from potential damage. The other purpose of the grounding system is to remove noise from the communication system itself.

There are several types of grounding systems that can be deployed at a base station, host terminal, and switching complex. The system used for any of these locations is either a single or multipoint ground system. The system used is driven by the equipment manufacturer and local codes, but the single point ground system is the most common in telecommunication systems.

A grounding system is often referred to in terms of ohmage, and the typical value is 5 ohms for a good ground. The ground for systems is usually obtained by using ground rods, driven into the earth near the site and then connected to the building ground system. Water pipes are also used with the necessary precautions, even though normally shunned, to ensure good ground integrity. Another method is the chemical bath solution.

Conversion Tables

Tables 8-7 and 8-8 show many of the standard conversions needed for various measurements encountered in an LMDS system. Often, the more common conversion issues deal with converting from metric to standard and vice versa.

Table 8-8 contains the standard temperature conversions needed for a wireless system.

Table 8-7

Standard distance
conversion

From	To	Multiply By
Meters	Feet	3.28
Feet	Meters	0.3048
Miles	Kilometers	1.609
Kilometers	Miles	0.6214
Kilometers	Feet	3281
Feet	Kilometers	0.0003408
Liters	Gallons	0.2642
Gallons	Liters	3.785
Rods	Feet	0.06061
Yards	Feet	3
Yards	Meters	1.094
Inches	Centimeters	2.54
Centimeters	Inches	0.3937
Feet	Centimeters	30.48
Centimeters	Feet	0.03281

Table 8-8

Standard
temperature
conversions

From	To	Multiply By
Fahrenheit	Kelvin	$(F + 459.67)/1.8$
Celsius	Fahrenheit	$(C * 9/5) + 32$
Fahrenheit	Celsius	$(F - 32) * 5/9$
Celsius	Kelvin	$C + 273.1$

References

Smith, Clint and Curt Gervelis. *Cellular System Design and Optimization*. New York, NY: McGraw-Hill, 1996.

Smith, Clint. *Practical Cellular and PCS Design*. New York, NY: McGraw-Hill, 1997.

Technical Organization

## Chapter 9

Introduction

This chapter addresses the topic of how a company may want to structure the technical organization for an LMDS/PMP system. The technical organization discussed is based on a full-scale LMDS service provider that could be a regional or a national organization. The way the technical organization is structured, including training, will directly influence the degree of success or failure that the communications company experiences. Head count drivers as a function of customers, host terminals and base stations will be put forth, including the use of smart outsourcing are also addressed.

Many wireless organizations begin centralized at their infancy and then move toward a more decentralized role, which is followed by the common issue of centralizing many functions again. The pendulum for centralization and decentralization tends to swing with each management change that the organization has. The reason why the change occurs is largely driven by an underlying problem that management attempts to correct by changing the organization structure.

It can be argued that if the organization has a plan from the beginning, which addresses the company's vision that matches the business plan, the need for dramatic organization changes will be minimized. If this is done, the organization problems that the company will experience will be largely driven by external influences (that is, the market condition), not internal pressures that result from its structure.

Figure 9-1 is an example of the organization structure that an LMDS operation can utilize. The structure as shown provides centralized control of the organization for common functions, but provides the market-sensitive organization needed in the ever-changing market climate.

The Technical Services group was singled out for the example. It can be responsible for the entire company's technical requirements, or there can be several VPs of Technical Services reporting to a Senior VP or directly to the President (depending on the span of control that is directly related to the amount of markets that the company has a presence in). The technical community is set up as a cost center, although the profit center issue is always proposed but not practical because the group does not have a direct revenue line, except for possible transfer costs. Because the technical community is a cost center, the head count drivers are largely driven by network components and not tied directly to revenue. However, sales is a profit center, an obvious issue, and the head count drivers for sales are based on revenue and not network components.

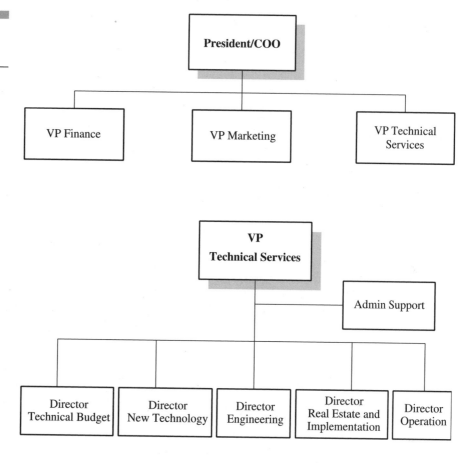

Figure 9-1
LMDS organization
structure

Technical Organization Structure

The technical organizational structure utilized by your company should be
driven by functional requirements. Functional requirements need to be the
driving force behind an organization's structure instead of personalities.
However, personalities often define the organization's structure because the
organization is arranged by who is in the group and not who should be
doing the work. See Figure 9-2.

The technical organization can take on a centralized role or a distributed
role for the company. It is recommended that a blended approach (that is,
the difference between centralized and decentralized) be utilized for the
company's technical structure. The concept of using a centralized versus a
decentralized approach is largely dependent upon the company's vision,
business plan, and culture.

Figure 9-2
Technical
organization
structure

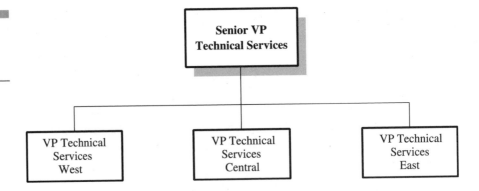

The centralized approach has the advantage of potentially achieving economies of scale, which are achieved through the elimination of redundant functions in each of the markets or areas. An example of a centralized function is new technology research for the company. New technology research for the company does not have to be accomplished with every department or division within a company. The idea of having one group lead the effort will ensure uniformity, accountability, and the probability that the direction picked is coordinated between the various organizations in the company.

The centralized approach, however, has the disadvantage of being defocused on the market requirements. The defocusing of market requirements can come about through not having any local knowledge of the technical configuration for the network. An example of defocusing can occur over a simple matter of switch port assignments, or services actually needed to be offered.

The decentralized approach has the advantage of being more market-sensitive and flexible than the centralized approach. An example of the flexibility to the market environment would involve the continuous configuration of the network, based on TDM and IP traffic patterns. The idea of the decentralized approach is that the decisions that will affect the market are brought as close to the customer as possible.

The disadvantage with the decentralized approach is the amount of redundant work that is performed. The decentralized approach lends itself to localized procedures that foster inefficiencies and a lack of knowledge transport. The lack of knowledge transport often leads to the problem or situation being repeated in another market, when some simple communication could convey how it could possibly be avoided. The decentralized approach also does not lend itself to any engineering practice procedures, which is essential in the rapidly changing world of wireless communications.

However, using a straight centralized or decentralized approach is not necessarily the best one. The organizational structure for the technical organization should take on a blended approach to centralization and decentralization.

The technical organization can take on a few variants in the approach, which can either be envisioned from the start or due to *(Mergers & Acquisitions)* M&As. Specifically, the variants to the organization pertain to whether the company has several divisions or does not have any other divisions. If a company has another division (for example, a west coast operation and an east cost operation), each would be considered a different market and would have its own technical directorate.

Regardless of which approach is taken toward the organizational structure, the span of control for each level of management should not go beyond seven or eight for any organization because of difficulties in managing such an organization for any prolonged period of time.

The technical organization comprising engineering, construction (implementation), and operations should be structured based on functional requirements. Functional requirements need to be the driving force behind an organization's structure instead of currently available personnel.

The technical organization should utilize a blended organization approach, taking advantage of both centralized and decentralized functions. The structure should be such that centralization occurs at corporate headquarters while locality to the markets is done through a distributed method.

Regardless of whether a decentralized or centralized approach is taken toward the organization structure, the span of control for each level of management should not go beyond seven or eight for any organization due to difficulties in managing such an organization for any prolonged period of time.

Technical Organization Departments

The key roles listed in the proposed technical organization, as shown in Figure 9-3, involve a multitude of disciplines, which comprise a well-rounded organization. The disciplines needed for the organization are as follows:

1. Technical Directorates
2. New Technology
3. Budgetary

Figure 9-3
VP technical services
organization

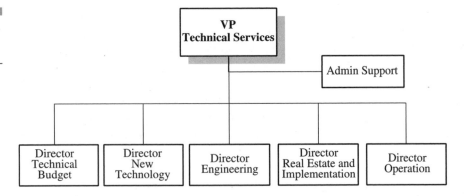

The key roles listed in the proposed technical organization involve budget, new technology, engineering, real estate and implementation, and operations directorates. The general functions of each of the directorates are as follows:

1. Budget is responsible for capital and expense tracking, variance reporting, forecasting, and purchase order handling for the entire technical organization.

2. Engineering is responsible for the design of the network and the technical performance aspects.

3. Real estate and implementation is responsible for the acquisition, leases, civil work, and construction of the various projects put forth by engineering.

4. Operations is responsible for ensuring that the equipment installed in the network is maintained and operating at its peak performance. The operations directorate is also responsible for the operation of the *Network Operations Center* (NOC), which can consist of a singular location or multiple NOCs.

5. New Technology is responsible for pursuing the future needs of engineering, operations, implementation, and marketing organization requirements.

Engineering

The engineering department is responsible for the planning and design of the LMDS communication system from a RF and network perspective. The

Figure 9-4
*LMDS engineering
department structure*

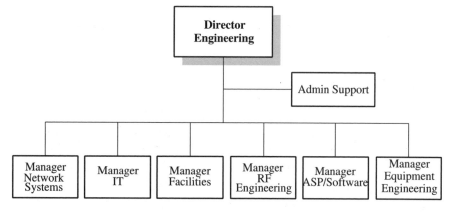

organization structure shown in 9-4 represents an organization that can be used to support either an individual market or a region that includes several markets.

The organization for engineering shown in the figure can be based on a world-wide level, regional level, country level, or even city level. The structure that can be followed is basically the same, no matter how many systems or sub-systems the organization needs to be responsible for.

The network systems group is responsible for the architectural engineering of the network growth, as well as evaluating new network designs and performing network troubleshooting. The network system group plans the switch dimensioning and module growth, and decides when a new switch is needed for the network with its proposed location. The group also is responsible for the SS7/CC7 network, as well as the layout and functionality of the IP and ATM network layers.

The IT group is responsible for the LAN/WAN configurations that are utilized throughout the corporation. The IT group is traditionally located within its own organization, but with the convergence of IP, ATM, and telephony, the IP program that is utilized and the associated platforms need to be put under engineering control.

The facilities management group is responsible for all aspects associated with facilities and utilities management. The group is responsible for ensuring that the trunk designs for the network are adequate and properly dimensioned for growth. Another role the group has is ordering the actual facilities for the network and ensuring that they arrive at the predetermined time for the project. The group is also responsible for dimensioning the DACs in the network and interfacing all the issues associated with the IXEs.

The RF Engineering group is responsible for the base station and host terminal planning and design efforts, including the PtP links. The group also is responsible for the frequency management of the network, intersystem coordination, regulatory, and aeronautical compliance for new sites. The group is also responsible for the overall performance of the existing RF network of the system.

The manager for ASP/software is responsible for all the application-specific software that is utilized in the system and by the host terminal equipment. The group ensures uniformity of applications for ease of expansion, troubleshooting, and cost savings. The ASP/software group is responsible for leading any software FOA for the switches and cell sites. The group oversees all major feature introductions into the network. The group also produces all the data translations information that is used for the switches base stations and SLAs. Another aspect the group is responsible for is the validation of the billing system whenever there is a software load change for the cell sites or switches. This group also performs system software audits to ensure that the software and translations are consistent across multiple nodes in the network, and it removes all unwanted data.

Equipment engineering is responsible for the physical layout and equipment requirements for the nodes, host terminal, switching, ATM, IP, and ancillary platforms. They are responsible for ensuring configuration management in the network.

Figure 9-5 shows levels within the engineering directorate. Naturally, some of the functions may not be necessary, depending on the services offered, for example, circuit-switching if only IP services are offered.

NOTE: *(x) is head count driver found in the section titled "Head Count Drivers."*

Operations

The operations directorate is responsible for managing the NOC; maintaining the RF, IP, ATM and circuit switch components; and managing the host terminal. The operations directorate will also be responsible for all host terminal upgrades that take place, some of which require physical visits. See Figure 9-6.

The NOC Manager is responsible for operating and maintaining the network operations center for the LMDS system which includes problem resolution, maintenance scheduling and technical support of the operations

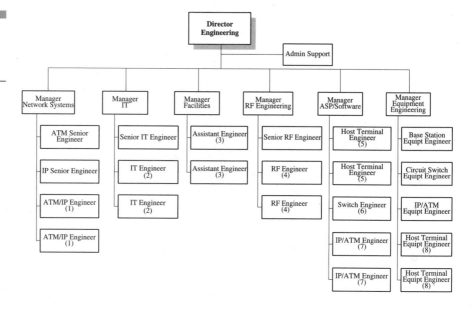

Figure 9-5
Expanded engineering organization

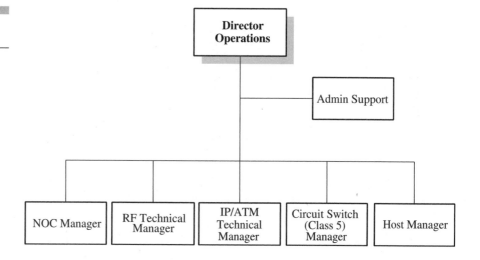

Figure 9-6
Operations

department. The RF technician manager will be responsible for the operation and maintenance the node sites, the host terminal, and as many microwave backbone systems that are put in place.

The IP/ATM tech manager will be responsible for the operation, maintenance, and implementation of engineering work orders for those platforms.

The class 5 manager will be responsible for the operation, maintenance, and implementation of engineering work orders for those platforms.

The host terminal installation manager will be responsible for upgrading the host terminals at the customer facilities. See Figure 9-7.

Real Estate and Implementation

The real estate and implementation directorate is responsible for securing, constructing, and commissioning the designs put forth by engineering. The groups directly in the directorate are the site acquisition, lease management, implementation manager, and bid manager. See Figure 9-8.

The site acquisition manager is responsible for obtaining both the nodes and switch concentration locations as securing access for host terminal installations at buildings. The lease manager is responsible for managing the multitude of leases that will exist as the system grows. The implementation manager is responsible for physically constructing the various facilities and commissioning them for conformance to engineering guidelines. The bid manager's job is to issue, manage, and award the numerous bids that will be issued to subcontractors and vendors.

Figure 9-7
Operations
organization

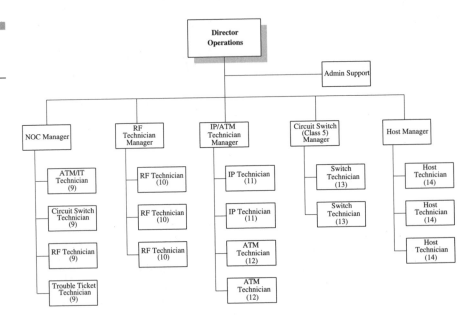

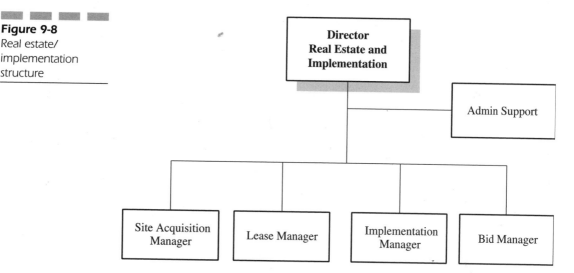

Figure 9-8
*Real estate/
implementation
structure*

The next layer down in the real estate and implementation directorate is shown in Figure 9-9.

New Technology and Budget Directorates

Figure 9-10 is an example of the directorates for new technology and budget control with an LMDS system.

The budgetary roles involve tracking, variance resolution and forecasting the capital and expense budgets for each of the various groups under this organization. The budgetary group in the organization plays a vital role in helping secure the necessary funding for various projects.

The budget staff listed in the chart is responsible for tracking, variance reporting, forecasting, and purchase order handling for the entire technical organization. The New Technology Director is responsible for establishing the technical vision of the technical community. Some of the roles the group perform, besides new technology investigations is being the central clearinghouse for technical performance requirements leading to a standard set of criteria for all to use. This group also ensures that a best practice approach is accomplished between all the markets, allowing for good ideas and procedures to be shared.

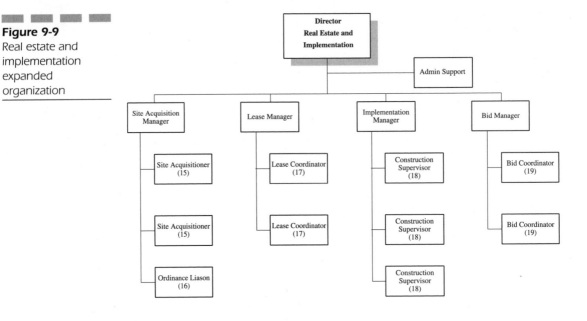

Figure 9-9
Real estate and
implementation
expanded
organization

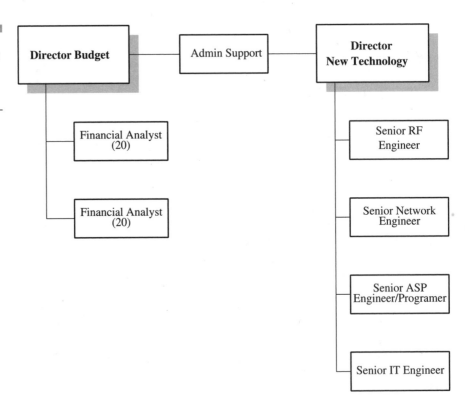

Figure 9-10
New technology and
technical budget
organization
structure

Head Count Drivers

Head count drivers, or ratios, should be used to dimension any organization and especially the technical community within an LMDS system. The drivers will need to be reviewed on a regular basis with the advent of new technology, better management systems, service changes, and the market conditions that dictate cash flow for the organization.

Ordinarily, the head count driver determines how large an organization should be as a function of success because the other direction is not a desired alternative. The goal is to determine when and where the hiring needs to take place. Obviously, additional burdens are placed on the organization with every new hire—the administration, space and support material they need for effectively doing the job they were hired to perform.

The personnel drivers for technical head count involve the following major items:

Number of customers

Traffic volume

Services offered

Number of base stations

Number of host terminals

This list is not all-inclusive but it serves as a good starting point from which to determine your staffing requirements. Revenue, of course, is a major driver for the LMDS organization as a whole, but the technical organization is typically a cost center and not a revenue center. Because the technical organization is primarily a cost center, the drivers shown are cost or rather function oriented.

Just what are the particular head count drivers? When referring to the various organization structures shown in Figures 9-5, 9-7, 9-9, and 9-10, the following drivers can be used. Obviously, the drivers are dependant upon the five basic items listed previously, but the example that follows will help structure the formulation of the drivers and the subsequent organization structure.

The following drivers shown in Tables 9-1, 9-2, and 9-3 can be used to dimension the initial and future organization. The head count data then can be entered into the operating expense part of the budget. Of course, some of the head count (associated with design and implementation) should be allocated to the capital budget.

Hiring

There are several methods that are used for securing the personnel for an organization. The paths for hiring can take several directions: from relying on the entire internal staffing to that of almost complete outsourcing. But no matter which method or variant to the staffing methods is chosen, hiring will need to take place in order to ensure that the finite resources are properly allocated.

Table 9-1

Engineering head count

Driver Number	Title	Dimensioning (per)
1	ATM/IP engineer	500 VPI/VCIs
2	IT engineer	50 host terminals
3	Assistant engineer	250 leased lines
4	Senior RF engineer	15 base stations and 100 host terminals
5	Host terminal engineer	200 host terminals
6	Switch engineer	200 ports
7	IP/ATM engineer	100 host terminals
8	Host terminal equipment engineer	250 host terminals

Table 9-2

Operations head count drivers

Driver Number	Title	Dimensioning (per)
9	NOC technicians	1.25 per shift
10	RF technicians	15 base stations and 40 host terminals
11	IP technicians	30 host terminals and 40 internal company workstations
12	ATM technicians	100 host terminals
13	Switch technicians	100 ports
14	Host technicians	75 host terminals

Table 9-3	Driver Number	Title	Dimensioning (per)
Real estate and implementation head count drivers	15	Site acquisitioner	50 base stations builds year
	16	Ordinance liaison	per market
	17	Lease coordinator	200 base stations and 500 host terminals
	18	Construction supervisor	25 base station builds year
	19	Bid coordinator	per market

One method is to hire the top management; they in turn hire the next level down in the organization, and so forth. This has the advantage of true accountability with the performance of the organization with the senior management. However, the key drawbacks are the time and the expenses that the company will incur due to the hiring method of bringing in the chiefs before the workers.

An alternative approach is to hire the one senior person for the organization and then hire the immediate line level for workers. This has the advantage of keeping the organization lean in the startup phase and reducing the organization costs. However, a disadvantage to this approach is the desire for the initial hires to jockey for the next-level position because it is vacant.

The third approach is to hire the top-level positions within the technical organization and utilize an outsourcing program to fill the needed void quickly. The outsourcing will be transitioned to permanent hires as the positions fill by using a normal interview process. This approach is the most economical method for obtaining a working staff right away, knowing that some organization rifts may occur due to the lack of any management layers when new hires are brought on board.

The third approach is the recommended method for starting up an LMDS system, was well as handling rapid expansions, which are a direct result of success problems.

Smart Outsourcing

Just what is smart outsourcing? Smart outsourcing involves utilizing outplacement firms, commonly referred to as consultants, for the purpose of providing the necessary manpower to meet the short-term staffing goals of the company.

I refer to utilizing outsourcing as "smart outsourcing" only if there is a plan of how to manage this resource. Often, the outsourcing is handled in such a fashion that it is loosely managed or not managed at all.

The key to smart outsourcing is to utilize selected consulting and outplacement firms that you will want to establish a long-term relationship with. Smart outsourcing should be part of your overall organization plan every year; it just depends on the level of outsourcing you want to utilize. The amount of outsourcing is largely driven by constraints, either in building up a permanent staff or limitations in hiring the required personnel in the time at hand. Also, smart outsourcing can be used to keep the existing staff focused on ensuring that revenue (for example, the customers and network are being optimized instead of chasing new technology issues or generating RFPs).

Smart outsourcing should not be used as a crutch for lack of hiring of entry-level staff. If this is the case, the use of outsourcing may be misapplied. However, outsourcing can be used to fill the entry-level positions for a transitional period, which needs to have a closure date while permanent staff is sought.

Outsourced technical resources can augment the existing technical community within an LMDS operation in the following suggested areas:

Training

Design reviews

Special projects

RFP generation and analysis

Staffing shortages

Mentoring program

Smart outsourcing should be part of your organization plan, but it needs to be managed to ensure that the resource is best used to benefit the organization, both from a short-term and a long-term aspect.

Training

Technical training for any organization is a difficult task because there is never enough time to ensure that proper training of personnel takes place. This section has a list of proposed general topics that should be utilized by the various engineering departments to ensure that there is a minimum level of formal training for all the groups.

It is strongly recommended that every new and existing employee be assigned a mentor for guidance. With the demand for immediate results for any issue, the use of a mentor program as an augmentation to formal training will expedite the training of personnel. The selection of who is qualified to mentor individuals needs to be taken with extreme caution to ensure that bad design practices are not propagated to the new members of the organization.

The training proposed for the engineering organization is defined in three sections: management, RF, and network. The rationale behind not selecting individual training programs for each of the individual organizations listed is multifaceted. The first rationale focuses on the need to have a certain degree of cross-training between the major groups in engineering. The second rationale deals with the focuses on the potential variances and individual market requirements for every organization.

The training that takes place for all the members of a technical organization should involve around 80 hours of training per year. The two-week training commitment is necessary for any technical organization to remain current and ensure that its personnel are adequately trained. It is strongly suggested that you employ a training de-confliction system to ensure that everyone receives training in the year and at the same time reduces the level of personnel outages caused by training.

The proposed training de-conflicting schedule lists all the training expected to take place for the employee and the department as a whole for the entire year. It is recommended that the schedule de-confliction have an initial granularity of monthly as a minimum. Regarding the following training de-confliction schedule, it is also important to list the personnel vacation plans, conferences, and major project milestones along with the training.

Month	January-1999	February-1999, etc.
Employee 1		
Employee 2		

The following is a brief listing of suggested technical material that needs to be taken by various major subgroups within engineering. The list provided, however, does not include the courses required by your company, such as orientation and various diversity classes. The list provided also does not define the number of classroom hours needed for each topic, nor the repeat interval for technical material. It is important to remember that certain courses should be required to be retaken at regular intervals to ensure that a minimum level of competency is maintained at all times.

Management

Basic Wireless Communications

Budget Training

Digital Radio Design

Disaster Recovery

E-Commerce/Web Design

Fraud Management

General Management

Interconnect

Interview Techniques

IPv4 and IPv6

Network Design

Network Fundamentals

Operations and Maintenance

Packet Switch Design (IP and ATM)

Presentation Techniques

Project Management

Real Estate Acquisition

RF Design

SONET/SDH

SS7 (CC7)

Statistics Theory

Switch Architecture

Tariffs

Telephony Call Processing

Traffic Theory

VoIP

Network Engineering

AIN

ATM Network Design

Basic Wireless Communications

Call Processing Algorithms

Cellular Call Processing

DACS

Data and Voice Transport

Disaster Recovery

Network Maintenance

Network Services

Numbering Plans

PCM and ADPCM

Performance Troubleshooting

Presentation Techniques

Project Management

RF Design

E-Commerce/Web Design	SONET/SDH
Equipment Grounding	SS7 (CC7)
Fiber Optics	Statistics Theory
Interconnect	Switch Architecture
Interview Techniques	Switch Maintenance
ISDN (BRI/PRI)	Tariffs
LAN/WAN Topology	Traffic Theory
MFJ/Descent Decree	Translations Switch
Network Architecture	Voice Mail
Network Design	WPBX
Network Fundamentals	

RF Engineering

Antenna Theory	Network Design
Base Site Installation	Performance Troubleshooting
Base Site Maintenance	PMP Frequency Planning
Basic Wireless Communications	Presentation and Speaking Skills
Digital Radio Design	Presentation Techniques
Disaster Recovery	Project Management
EMF Compliance	Real Estate Acquisition
Grounding and Lightning Protection	RF Design
Interconnect	Statistics Theory
Interview Techniques	Switch Architecture
Microwave System Design	Traffic Theory
Network Architecture	WPBX

Operations

Base Site Installation	LAN/WAN Topology
Base Site Maintenance	Network Architecture
Basic Wireless Communications	Network Maintenance
Fiber Optics	Performance Troubleshooting
Grounding	Switch Maintenance

Every employee, whether new or old, in a department needs to have a training program. The training program is more of a development program, with the desired intention of helping improve the skill sets of the individual. In addition to improving the skill sets for the individual, the department as a whole will be improved though higher efficiency levels, obtained with the new skill sets.

When crafting and authorizing any training program, it is essential that the subject matter is relevant to their training program and job function.

However, when a new employee arrives, it is imperative that they be given some guidance to help ease the transition period. A recommended format for establishing a new-hire training program is listed as follows. The details for each section will need to be tailored for the particular job function, but the general gist of the program can be easily extracted here.

To: Name of Employee

From: Manager

Date:

Subject: Training Program for Position X

_____

Objective:

Define here what the stated objective is for the training program
    and its duration

Description of Job:

This is where a description of the actual job is specified.

Reports:

Description of the various reports needed to be generated by the
    individual, when they are due and to whom they should be sent.

Reading Material:

This is a brief description of the suggested reading material the
    person should have in their personal technical library. It is
    imperative in this section that the source locations for the various
    documents be spelled out.

Training Program:

This section involves putting together a training program for the
    individual. The mentor assigned to the person also needs to be
    defined.

Assigned Project:

This section lists the particular projects assigned to the individual and their deliverables, with a timeline.

References

AT&T. *Engineering and Operations in the Bell System*, 2nd Edition. AT&T Bell Laboratories, 1983.

Aidarous, Salah and Thomas Plevyak. *Telecommunications Network Management Into the 21st Century*. New York, NY: Institute of Electrical and Electronics Engineers, 1994.

Chorafas, Dimitris N. *Telephony: Today and Tomorrow*. Upper Saddle River, NJ: Prentice Hall, 1984.

Degarmo, E. Paul, William G. Sullivan, James A. Bontadelli, Elin Wicks. *Engineering Economy*, 11th ed. Upper Saddle River, NJ: Prentice Hall, 1999.

Keller, Gerald and Brian Warrack. *Statistics for Management and Economics*, 5th ed. Duxbury Press, 1999.

Smith, Clint and Curt Gervelis. *Cellular System Design and Optimization*. New York, NY: McGraw-Hill, 1996.

Performance Reports

Introduction

This chapter will discuss the various levels of documentation and system reports that need to be generated by the technical community within an LMDS system. The reports presented are recommended reports and are structured in a generic fashion for guidance in establishing the various types of reports that are required for any operating system. The reports are geared toward a full-service LMDS operator, but can easily be parsed down or added to, based on the particular structure of your company, the platforms used for service delivery, and the services themselves.

Reports referenced throughout this chapter relate to both written and verbal reporting. The objective with performance reports is to facilitate the proper management of the network as a whole, which will result in improved performance and utilization of the capital infrastructure. The reports, if structured properly, can and will be used to prevent and correct system-related problems. The reports should also address requirements for designing the network in support of sales initiatives. Also, the reports can also help facilitate the use of add-on selling by helping inform the sales staff of their clients' usage patterns when SLAs could be adjusted to better meet their usage.

Regardless of the type or function of the reports, a hierarchical approach to system report generation is strongly recommended. The rationale behind a hierarchical report approach pertains to the information requirements needed by different levels of management in an organization. Specifically, a report that is used directly by a network engineer should be different from the report issued to a vice president for the same topic. Although the hierarchical approach for report generation and dissemination seems straightforward, there are many reports that go directly up the chain of command in a raw format.

The examples of the system reports presented in this chapter include network performance, RF performance, exception reports, customer care reports, construction status, operations, software configuration, and engineering project reports.

It is suggested that each report have a description of the process of the way it was produced, attached, or immediately available. The description of the report process should include the following as a minimum:

- All the data contained in its report and the sources of the data
- All the equations used in the calculations and the field in which they are applied

■ A listing of the software used in the processing of the data

It is also suggested that each report be checked for accuracy by running a set of simple values through the processing algorithm. The sample data that is run through the processing algorithm should also be checked against a manual run of the same data by using a calculator or slide rule. Although this seems like a rather basic issue of checking the reports themselves, often the reports themselves are never checked, resulting in numerous errors that occur as a result of misinformation.

There are some basic concepts to report generation and dissemination that you need to address for every report you are currently generating. When addressing any report or documentation process, it is important to obtain the answers to the follow seven questions:

■ What is its purpose?

■ Who will generate the report on an ongoing basis?

■ Who will act on the information that is in the report?

■ Who will receive the report?

■ Is the report needed?

■ What format should the report be in (that is, electronic or paper)?

■ How will the report be processed?

If reporting processes have not been fully implemented, it is strongly recommended that you identify all the reports currently being generated in the various departments within engineering alone. A simple review of all the reports that are being generated will most likely result in several of the reports being identified as no longer being needed. The review of the reports will also point to duplications in efforts within the organization itself. The duplication of various reports can be eliminated and resources can be better utilized to focus on other topics of direct interest to the company.

The reports themselves should utilize a document control number and be stored in a central location. The central location recommended for depositing all the engineering documents is the engineering library.

The engineering library should contain all the meeting and/or project notes and all system reports. To facilitate access to the information, the meeting and/or project notes along with all system reports should also be resident on the engineering LAN. Therefore, the engineering library consists of both electronic and physical media. There is a need for both physical and electronic media due to the method that information is sent to the technical community. Not all vendors present their materials in a CD-ROM, for example, and trade magazines are still predominantly physical at this time.

If an engineering library does not exist in your company, it is strongly suggested that you develop one. The departmental administrator or a separate assistant should be placed in charge of the engineering library. The individual in charge should be responsible for overseeing the signing-out and return of manuals, reports, etc. The administrator would also have the responsibility of updating the library and discarding outdated material. Examples of materials to have in the engineering library include industry specifications, current vendor manuals and reports, and copies of RFP responses.

It is as important to ensure that there is sufficient information in the engineering library as it is to prevent unwanted material from being stored there. The engineering library physical media location should not be a technical dumping ground for industry magazines, software manuals, and other items collected by the engineering staff.

The same issue goes for the electronic media; the file structure used should be organized in a logical fashion to facilitate retrieval of information. The electronic storage privileges should be restricted to a few key personnel who have the permissions to add files to the engineering library. Everyone else should have read-only privileges.

The success or failure of the engineering library is totally dependent upon continued vigilance to ensure that current and relevant information is stored in the facility, and that is is accessible only by the groups requiring the information; not everyone.

Reports

What are the relevant reports that need to be looked at by the line personnel responsible for maintaining and dimensioning the network? In addition, what are the types of reports that management needs to base their resource allocation decisions upon? The answer, as expected, is not simple or just overly complicated. Fundamentally, you need to have information on every element in the system and performance data that is relevant to those platforms. The performance information relative to the platforms needs to be at a sufficient level so you are ensuring that the services offered are being provided at or exceeding the particular SLAs that are offered. The performance information is also meant to help dimension the network elements by providing the information that says how the network is handling the various types of services with their relevant traffic loads.

Just what are the types of reports that need to be generated and reviewed for potential action? The general categories are listed as follows for quick reference. The list, of course, can be added to, but the key concept to factor into the reporting process is to decide what you need to respond to key actionable time lines: daily, weekly, monthly, quarterly, or yearly, for example.

- Bouncing congestion hour traffic report (node and service)
- RF network performance reports
- Packet-switch performance report
- Circuit-switch/node performance report
- Telephone number inventory report
- IP number inventory report
- Facility usage/traffic report
- Facilities interconnect report (data)
- System traffic forecast report
- Network configuration report
- System growth status report
- Exception report (morning, weekly)
- Customer care report
- Project status report (current and pending)
- System software report
- Upper management report
- Company meetings
- Network briefings

The list also can include sub-categories where additional granularity is achieved. However, it is important to always keep in mind that reports need to provide actionable information that is relevant to providing the best quality network while minimizing capital and operating costs.

Bouncing Congestion Hour (BCH) Traffic Report

The *bouncing congestion hour* (BCH) traffic report's intention is to identify the system and individual node busy hours. Through identifying system and individual node sites, the busy hours of any network or local congestion problems can be identified. The BCH report should be generated on a biweekly and monthly basis. The reporting interval is such that the data to respond to is design-oriented and not operational-related, which is captured in the exception report. The BCH report also needs to include weekend data and the weekday traffic information for a 24-hour period.

This level of information should be disseminated only to the network and RF engineering for the network on a regular basis.

The system-level report should be presented in both tabular and graphic method for ease of digestion. The information should be broken down to represent individual days of the month and compared to the previous year(s) data for trending information, when the system is more mature, or on a monthly basis for initial system inception.

Table 10-1 is an example of a BCH report format. As with all reports, the specific infrastructure of the network and the services offered need to be included in the decision about what needs to be monitored. For instance, if a best effort for IP traffic is all that is offered, monitoring the congestion level for the Internet will not be as important as if video conferencing is being offered as a service.

The biweekly report can also be used as a sales-support tool, which is where the individual customer's traffic is monitored and compared to the SLA they have signed up for. Based on the usage levels coming from the system, the information should be discussed with the customer to ensure that the service they have chosen with sales support was correct. In addition, this can be a chance to possibly identify a configuration problem that the customer may have, in which they may be inefficiently using the service, resulting in a higher cost than what is needed.

The data in the table is straightforward and requires little explanation. However, it is not recommended that any averaging method be utilized for these particular reports (that is, weekly). Instead, the individual bouncing congestion hour data for each element needs to be reported.

As mentioned previously, the reports should only be disseminated to the network and RF engineering groups. The same report should, however, be distributed to the technical director for network and RF on a monthly basis.

Table 10-1

Bouncing
Congestion Hour
(BCH) report

								Packet Platforms			
Node	Time	Port Number	Service	CIR	(Mbps)	Mbps Avg.	Mbps Peak	Band-width Available	Link Utiliz-ation %	Con-gestion (sec)	OOS (min-utes)
ATM											
Router (IP)											
Frame Relay											

			Circuit Switch		
Switch Number	Time	Port Number	BH Traffic	Blocking %	Utilization

			Base Station					
Station Number	Sector	Radio	Design Bandwidth (Mbps)	Avg. Mbps	Peak Mbps	Radio Utilization %	Congestion (sec)	OOS (sec)

				Customer			
Account Number	Contact	SLA	CIR	Avg. Mbps	Peak Mbps	Congestion (sec)	Network Availability (sec)

This report should not be distributed to higher levels of management because it is really meant for the working level to utilize.

The customer BCH report should be disseminated to the sales and marketing departments on a monthly basis. Obviously, more detail can be added to the customer BCH report.

RF Network Performance Report

The RF network performance reports vary, based on the infrastructure vendor used and the services offered. However, the proposed performance reports for an LMDS system are discussed next and relate to the many elements in a network that may or may not be in existence.

The performance reports should be available on a daily and weekly basis with the added benefit of being able to trend the performance over a period of time to spot trends. With each report, the specific order of the information and its particular format should be changed to facilitate ease of use and also functionality for identifying problems. See Figure 10-1.

Table 10-2 shows the specific performance reports related to the various RF network components depicted in Figure 10-1.

Packet-Switching Performance

The following list of performance reports relates to the packet- and circuit-switching aspects of an LMDS system. The components that comprise the network elements of an LMDS system differ by operator, but Table 10-3 provides sufficient detail from which to begin monitoring and refining the network. The description of reports assumes that there is an ATM switch along with a router, but other services are handled by another service provider.

Figure 10-1
RF network
components

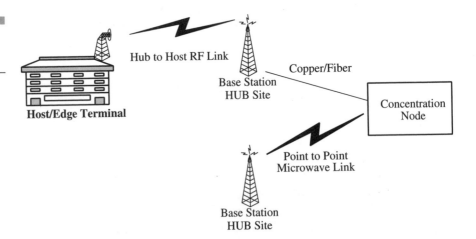

Table 10-2

RF network configuration report

Hub Site Number

Date:

Number of sectors (type of LMDS cell)

(2, 3, 4, 6, 8, 12 sector)

Number of radios/sector

Sector polarization (V, H, H/V)

Number of host terminals per sector

Number of customers per LMDS sector

Radio type (bandwidth)

Modulation technique

Maximum Mbps

Radio link utilization (average)

Peak link utilization (24 hr.)

Number of assigned service types (IP, LL, FR)

Available Bandwidth left (Mbps)

Outage (seconds) daily

Outage (minutes) weekly (rolling 7 days)

Host Terminal Number

Date:

LMDS cell assigned

LMDS radio/sector

Polarization (V, H, H/V)

Number of customers assigned to terminal

Central or distributed terminals

Radio type (bandwidth)

Modulation technique

Maximum Mbps

Radio link utilization (average)

Peak link utilization (24 hr.)

(continues)

Table 10-2

RF network
configuration
report (continued)

Number of assigned service types (IP, LL, FR)

Available bandwidth left (Mbps)

Ancillary platforms (routers, MUX, etc.)

Outage (seconds) daily

Outage (minutes) weekly (rolling 7 days)

Point to Point (Ptp) Links

PtP number

Channel

Tx frequency

Rx frequency

Link utilization

Link size

Peak utilization of link (24 hr. look)

Type of link multiplexed or home run

Mbps of usage

Protected/unprotected

Hub sites included

Outage (seconds) daily

Outage (minutes) weekly (rolling 7 days)

Hub to Switch—Link utilization

Link number

Link size

Peak utilization of link (24 hr. look)

Multiplexed or home run

Type of link

Mbps of usage

Protected/unprotected

Fiber/copper/PtP

Outage (seconds) daily

Outage (minutes) weekly (rolling 7 days)

Table 10-3

Packet-switching configuration report

ATM Switch
ATM switch number
Date
Type (edge/core)
Number ATM ports
Number IP ports
Number FR ports
Number CES ports
CPU utilization %
Number PNNI
Number PVC
Number SVC
% On-net traffic
Cell error rate
Severely errored cell block rate
Cell loss ratio
Cell misinsertion rate
Cell transfer delay
Mean cell transfer delay
Cell delay variation (CDV)
% AAL1
% AAL5

Router
Router Number
Date
Type
Number T1/E1 ports
Number OC3 ports
CPU utilization %

(continues)

Table 10-3

Packet-switching
configuration
report (continued)

Number PVC

Number SVC

Peak Mbps

Average Mbps

Cell latency

Max number simultaneous connections

Connection utilization %

Packet retransmission %

On-net traffic %

Inbound traffic %

Circuit Switch/Node Metrics Report

The switch/node metrics report is intended to assist in the monitoring and dimensioning of the circuit switches in the network. This report should be generated once a week for each switch/node in the system and be provided to the network engineering department for review. This report should include, as a minimum, the following data fields:

Performance Data:

	Recommended Sample Times:
- Central Processor (CPU) load/utilization value	System busy hour %
- Secondary processors (SPs) load/utilization value	System busy hour %
- Port capacity assigned	Weekly average
- Port capacity available	Actual
- Subscriber capacity assigned	Weekly average
- Subscriber capacity available	Weekly average
- Memory capacity assigned	Weekly average
- Memory capacity available	Weekly average
- Switch/node I/O capacity assigned	Weekly assignments
- Switch/node I/O capacity available	Actual
- Service circuits load/utilization values (for senders, receivers, tone generators, etc.)	System busy hour %
- Switch/node outages (number and duration of occurrences)	Daily recordings

As mentioned previously, this report should be generated once a week as part of the normal maintenance and operation of the network switches and nodes. However, if a problem is encountered or if any metric listed approaches an operational limit, then this report or the metric(s) of interest should be collected and reviewed on a daily basis. This will provide better accuracy when monitoring the switch or node of interest.

The proposed switch/node metrics report is shown in Table 10-4.

Table 10-4

Switch/node metrics report

Switch/Node Metrics Report	
System Name: _____	System "X"
Date: _____	
Report Week: _____	
Node :_____	Switch #1
CPU Load / Processor Occupancy :_____	47%
Secondary Processor #1 Load: _____	29%
Secondary Processor #2 Load :_____	20%
Secondary Processor #3 Load: _____	23%
Node Port Capacity: _____	985/1200 matrix ports
Node Subscriber Capacity: _____	45,000/70 ,000 subscriber records
Node Memory Capacity: _____	10M/35M
Node I/O Capacity: _____	12/21 I/O ports
Service Circuits Loading:	
Sender Circuits: _____	30 %
Receiver Circuits (MF / DTMF): _____	43%
Conference Circuits: _____	23%
Node Outage Data _____	No outage for this report period
Start time of outage: _____	
End time of outage: _____	
Duration of outage: _____	
Reason for outage: _____	

Telephone Number Inventory Report

The use of the telephone number inventory report is vital for monitoring the telephone numbers usage (that is, how many actual numbers are assigned to the customer base at a given time). The report also is used for predicting, projecting, and trending the directory inventory growth. Through knowing the telephone number growth patterns, an engineer can order additional codes for future use from the *Local Exchange Carrier* (LEC) or the governing body for code administration.

This report needs to be generated on a monthly basis and should indicate as a minimum the complete breakdown of all the directory numbers used in the network. The report also needs to include the actual network nodes where the numbers are stored. The telephone number inventory report should also indicate the status of future codes soon to be introduced in the network and when they are expected to be released. Additionally, the report needs to include the actual *central offices* (COs), where the codes are served out of, to ensure proper planning and expediting troubleshooting.

The telephone number inventory report should be distributed to the network engineer responsible for the inventory tracking and forecasting, the network engineering manager and the director of engineering.

An example of the telephone number inventory report is shown in Table 10-5.

IP Number Inventory Report

The use of the IP addresses and their related subnets is essential for any LMDS system. The IP addresses are one of the most prevalent forms of identification for all elements within the LMDS system and are also a vital service supplied to the customers. There are both public and private addresses that need to be tracked. The IP number inventory report is vital for monitoring the IP address usage (that is, how many and how they are distributed at a given time).

The report also is used for predicting, projecting, and trending the IP inventory depletion and subsequent growth. Through knowing the IP address growth patterns, network engineering can order additional public addresses when needed or reserve the correct number of private addresses for future use.

Telephone Number Inventory Report

System Name: _____

Week Ending: _____

Date: _____

NPA	NXX	XXXX	Available	Central Office	Resident Switch	Route Name	Tandem	Tested	Number Released	Number Active Released	Utilization Rate	Comments
914	365		7000–7999	8/15/95	ORB	WDB1/003	BBEN9	ZERK	1000	1000	500	50%	*
201	968		2000–2999	8/26/95	PARM	ERU07/011	CPG2/4	NWK	1000	500	200	20%	

Table 10-5 Sample telephone number inventory report

This report needs to be generated on a monthly basis and should indicate as a minimum the complete breakdown of all IP addresses used in the network. The IP number inventory report should also indicate the status of future addresses soon to be introduced in the network and when they are expected to be released. Additionally, the report needs to include the actual elements or nodes where the codes are served out of to ensure proper planning and expediting troubleshooting.

The IP number inventory report should be distributed to the network engineer responsible for the inventory tracking and forecasting, the network engineering manager and the director of engineering.

An example of the IP number inventory report is shown in Table 10-6.

This report format is crude, but the general concept of reporting can be achieved by defining which elements in the network require IP codes, those that are private and public, and their distribution or assignment method.

Facility Usage/Traffic Report

The facility usage report's purpose is to track the various interconnect facility usage levels in the network. The report should be issued on a monthly basis and used for many issues associated with improving the network's facility performance.

The facility usage report should be used to determine the best locations for the *point of presence* (POP) locations in the network. The selection of the POP locations, if done properly, will minimize the network infrastructure cost for delivering calls.

The report needs to also be used for verifying that the interconnect bills received for operating the network are valid. The interconnect bills should be reconciled against the facility usage report to ensure that there is nothing out of the ordinary being reported or billed. The reconciliation of the facility usage bill is normally the responsibility of a revenue assurance department.

The facility usage report should be distributed to the network facilities engineer responsible for facility usage dimensioning, the network manager and the director of engineering, who should forward a sanitized version of the facility usage report to upper management with the interconnect bills.

An example of what a facility usage report should look like is shown in Table 10-7.

IP Number Inventory Report

System:

Date:

Public Addresses	Node	IP Block Assigned	Subnet Range	% Utilized	Customer	IP Address	Subnet
	A						
	B						
	C						

Private Addresses	Node	IP Block Assigned	Subnet Range	% Utilized	IP Address	Subnet	Max IP Terminals Element	Customer
	A							
	B							
	C							

Table 10-6 IP number inventory report

Table 10-7

Sample facility
usage report

Facility Usage Report						
System Name						
Month						
Date						
Trunk Number	Usage	Band	Band Usage %	Variation % Previous Month	Design %	
NJ004-05	8545	1	25%	4%	25%	
		2	35%	1%	35%	
		3	30%	5%	25%	
		4	10%	6%	15%	

Facilities Interconnect Report (Data)

The facilities interconnect report (data) is intended to display the current configuration and performance of the network data links for detecting and resolving data problems, and to plan for the dimensioning of these facilities as the network traffic grows. These are not the switch-to-cell-site links, but instead the switch-to-switch and/or switch-to-network node links responsible for call processing and database inquiries.

One of the most prevalent types of inter-switch and inter-node interconnection protocols used in the cellular industry is the ANSI standard *System Signaling* #7 (SS7). We will assume these types of data links for this report. Other such transmission protocol types will use similar metrics to measure data link performance, such as those used for IP, Frame Relay, and ATM. Thus, for evaluating link performance, this report should include, as a minimum, the data fields listed as follows.

Basic Link Configuration Data:

- Defined network point codes
- Number of defined link sets in the network
- Number of links defined in each link set

Performance Data:

- Link traffic load (erlangs)   System busy hour data only
- Link set traffic load (erlangs)   System busy hour data only
- System traffic load (erlangs)   System busy hour data only
- Link active and inactivity times   Total peak service hours (0700–2000)
- Link set change over count   Total peak service hours (0700–2000)
- Link retransmission %   Total peak service hours (0700–2000)

This data can be collected by either using the network switches or by using separate monitoring equipment such as a protocol analyzer patched into the network links in a non-intrusive (passive) manner.

This report should be generated on a weekly basis, with the responsibility of its review assigned to the data engineering group within the network engineering department. If a problem is noticed on a link set or a link, however, a more detailed report about this problem facility should be obtained to conduct more in-depth troubleshooting.

The actual link/link set performance thresholds settings set should be based upon the ANSI standards manual to ensure accuracy and uniformity.

A sample facilities interconnect report format is shown in Table 10-8 for reference.

Network Configuration Report

The network configuration report is a collection of diagrams showing the network configuration as it exists. The network configuration report needs to be updated upon every major network change (that is, new switch, node additions, new central office exchanges, etc.). The network configuration report should be distributed to all of network engineering and the operations department as well.

The network configuration required tables and diagrams are as follows:

1. Voice interconnect diagram

–Concentration node to PSTN LEC central offices and IXC tandems
–Internal system voice facilities for call delivery

Table 10-8

Facilities
interconnect report

	Facilities Interconnect Report	

System Name: System "X"

Date:

Network Point Code Assignments:

Node Name	-	255 - 1 - 1
Node Name	-	255 - 1 - 2

Network Link Definitions:

Link Set: 255 - 1 - 1	Links: SLC - 1
	SLC - 2
Link Set: 255 - 1 - 2	Links: SLC - 1

Link Traffic Data:

Link Set 255 - 1 - 1	0.38 erlangs	SLC - 1 0.20 erlangs
		SLC - 2 0.18 erlangs
Link Set 255 - 1 - 2	0.15 erlangs	SLC - 1 0.15 erlangs

System Data Link Traffic Load: 0.43 erlangs

Link Service Data:

Link Set 225 - 1 - 1	Links: SLC - 1	No outages
	SLC - 2	0900–1000 hour, 02:00 min
	Link Change Over Count:	2
	Link Routing Error Count:	0
	Link Retransmission %:	SLC - 1 0.15 %
		SLC - 2 0.01%
Link Set 225 - 1 - 2	Links: SLC - 1	No outages
	Link Change Over Count:	0
	Link Routing Error Count:	0
	Link Retransmission %:	SLC - 1 0.01%

2. Data network diagram

   –IP links

   –Frame Relay links

   –ATM links

   –SS7 data links

3. Concentration node to base station assignments

4. Point to point microwave links

5. Host terminal to base station assignment

6. Auxiliary system interconnection diagrams

–Servers

A sample of a network configuration report is shown in Figure 10-2.

Exception Report

The exception report is one of the most valuable tools available for troubleshooting on a daily basis. The exception report should be generated by operations and engineering combined. The exception report should contain the basic information regarding the network from a maintenance point of view. At a minimum, the report should be distributed to the network manager, RF performance manager, and the operations managers. The directors for both engineering and operations would also benefit from seeing the report.

The one primary problem that will occur with this effort is the coordination efforts. It might be best to have one group be directly responsible for the generation of the report and have a key contact in the other group provide tactical data.

The exception report is meant to help identify the current level of maintenance issues in the network at any given time. The intent of this report is to help improve the troubleshooting response for engineering. The format of the report is shown in Table 10-9 for reference.

Figure 10-2
Network
configuration report

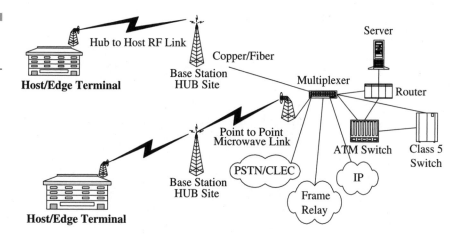

The Morning Report

Date: 1/1 /2000

Previous Days Bouncing Congestion Hour Data

Total IP Packets Transported (Mbps) 100000

% Retransmission   5%

Maximum Number Simultaneous Connections   500

Peak Mpbs   1250

Avg Mbps   500

Base Site Outage   1

T1/E1-Outage   2

Radios OOS   5

Host Terminals OOS   1

Planned Outages: (description)

Un-Planned Outages: (description)

The information in this report is time-sensitive and its dissemination is critical to the parties that need it. It is strongly suggested that if there is an electronic mail system used by the system operator, this report should be stored in a central location for quick reference.

Customer Care Report

Customer care reports are one of the key metrics for receiving information about the quality of the network. The customer care report should be integrated with the trouble reporting and resolution system that is used by customer care and engineering.

The frequency of the customer care report should be biweekly and it should be distributed to all the managers in engineering and operations. The level of detail and information content in the report is shown in Table 10-10.

Table 10-10

Customer care
report

		Region 1	Region 2	Region 3	System
Customer Care Report					
System Name:					
Date:					
Network Complaints		11	12	17	40
# IP-Related		3	2	5	10
# Frame Relay Complaints		5	9	5	19
# Circuit Switching Complaints		3	1	7	11
# Trouble Reports Issued		4	6	5	15
# Trouble Reports Closed		3	6	5	14
# Outstanding Trouble Reports		1	0	0	1

Project Status Report (Current and Pending)

The project status report is meant to track all major projects that are currently underway or proposed by engineering for the coming year. The report should have a rolling one-year projection as the best possible position to report on.

The report is meant to help identify major network activities that are current or will take place. The report should match the organization's goals and objectives. This report needs to be generated and issued on a biweekly basis to all the managers for engineering, operations and implementation. The objective is to ensure that all the technical community for the company is fully aware of all the projects engineering is currently or will be involved with.

The format for the report is shown in Table 10-11.

When the project is formed, the project plan should include an executive summary, objective, project prime, review dates, milestones, manpower loading, budget impact, impact to other departments, pass/fail criteria, and an MOP. When the project is completed, the post-implementation review and report needs to be generated by the project leader.

				Project Status Report				

System Name

Date:

Dept	Project Name	Priority	Originator	Lead Dept/ Person	Due Date	Project Plan	Capital Funding	Comments
RF Eng	Alpha	1	Marketing	RF/Smith	8/15 /00	7/16 /00	7/17 /00	On-target
Netwrk	Tree	2	Engineering	Netwk/ Gervelis	8/15 /00	7/20 /00	7/24 /00	On-Target

The executive summary for the project plan and the post-analysis can easily be used for conveying the objective and final outcome of the project to upper management.

The general format recommended to be followed is illustrated as follows.

Project Plan Report Format

1. Executive Summary (one page)

1.1 Objective

1.2 Expected Time

1.3 Manpower and Infrastrucute Requirements

1.4 Projected Cost

1.5 Positive Network Impact

1.6 Negative Network Impact

2. Project Desciption

2.1 Project Leader

2.2 Project Team

2.3 Project Milestones

2.4 Related Documents

3. Design Criteria

3.1 Basic Design Criteria

3.2 Project Review Dates

3.3 Method of Procedure

3.4 Hardware Changes

System Software Report

The system software report is another essential report. The purpose of the report is to identify to all the managers and relevant engineers in engineering and operations the current software configuration for the network. The report should be only one page in length and issued on a biweekly or monthly basis. See Table 10-12.

Naturally, if there are more than one vendor or type of device located in the network, then the list can and should be expanded upon.

Table 10-12

System software report

System Software Report			
System "X"			
Date:			
	Current	Tested	Next Load and Expected Date
Class 5 Switch			
Class 5 Switch CPU			
Class 5 Switch Matrix			
Class 5 Switch Database			
Router			
ATM Switch			
Voice Mail			
DXX			
PtP Microwave			
Hub Station			
Host Terminal			

Upper Management Report

The upper management reports are an essential element in an LMDS system's operation. The use of a series of quick and concise information to upper management will palate its thrust for critical knowledge on the performance of the network. The amount and types of reports that can be sent to the upper management of any company range for voluminous to sparse. It is recommended that the following information be produced and disseminated to all the management in the technical area of the company on a monthly basis.

The information recommended to be issued upward contains data extracted from most of the reports recommended to be generated in this chapter. The difference in the reports lies mainly in the format and amount of information content delivered.

The proposed upper management report is shown in Table 10-13.

Table 10-13

Upper
management
report

Upper Management Report
System Name
Date
System Measurements
Total Voice Minutes
Total Mbps Transported (data and voice)
Total IP Mbps
Total Frame Relay Mbps
Total ATM Mbps
Peak Mbps Transported
Average Mbps Transported
Average Customer/Host Terminal
Average Mbps/Customer
Average Customer/Base Station
Network Performance
RF System Availability
Backhaul System Availability
Switching Network Availability
Packet Network Availability
Overall System Availability
Planned Outage Time (minutes)
UnPlanned Outage Time (seconds)
Configuration
Host Terminals
Base Stations
PMP Radios (V)
PMP Radios (H)
Ptp Radios
PSTN Trunks (DS3)
IP Trunks (DS3)
Frame Relay Trunks (DS3)
ATM Trunks (DS3)
DXX
Routers
ATM Switches
Servers
E-Mail Servers
E-Commerce Servers
Web Hosting Servers

Company Meetings

Meetings are an essential form of report generation. The types and frequency of the actual meetings are indicative of the focus placed on designing and improving the network. The critical point is that there needs to be an effective balance between the number of meetings held and the need to convey information in a person-to-person environment.

It is recommended that you review the current meeting levels you are currently using and see if you are either meeting too frequently or not enough.

A suggested meeting structure is proposed for various levels in the organization:

Director to director	Once a week
Director to manager	Once a week
Manager to manager	Once a week (you need to talk with your counterparts on more frequent basis)
Engineering and marketing	Once a month
Engineering department meetings	Once a month
Engineering and customer care	Once a month
Engineering design reviews	Once a week
Project approval meeting	Biweekly
Network growth plan	Every three months

Except for the engineering design reviews, project approval and network growth plan, the meetings should not last more than one hour in length. It is recommended that the weekly status meeting between the groups take place at the beginning of the week instead of the end of the week. An agenda for each of the meetings should also be provided prior to the meeting so they stay focused.

Regarding company meetings: It is imperative that the marketing department should plan initial meetings with the engineering department prior to the onset of any new product or service. The objective with the meeting between marketing and engineering prior to any new product or service offering is to assess the feasibility of the produce or service offering. The meetings between marketing and engineering can take place more frequently than once a month.

The main and auxiliary system capabilities should be discussed with marketing as well to ensure that these system capacities can support the projected growth. The marketing departments need to provide for any new product and services they plan to launch to the customer base.

Regarding company meetings: It is important that the following key elements, or rules, be followed to ensure that they are successful. A meeting's success is defined as having met its stated purpose with a decision being reached or the required follow-up action items being defined and delegated.

Pre-Meeting Steps:

1. Plan what the meeting is about and its desired outcome.
2. Identify who should attend the meeting.
3. Develop the meeting agenda and distribute it in advance of the meeting.
4. Ensure that there is sufficient room and audio/video equipment for the meeting.

Meeting Steps:

1. Introduce everyone and clarify everyone's role in the meeting.
2. Review the agenda.
3. Cover each agenda item one at a time.
4. Allow for sufficient feedback on each topic and control unfocused people.
5. Close all discussions on the topic at hand before moving to the next agenda item.
6. Summarize all decisions and agree upon action items.
7. Draft agenda for next meeting and agree upon a time.
8. Close meeting by thanking everyone for attending.

Post-Meeting Steps:

1. Write and distribute meeting minutes promptly.
2. File agenda, minutes, and other key documents in the engineering library and LAN.
3. Follow up on all open items to ensure closure.

Network Briefings

Network briefings are an important aspect for reporting the status of the network (or state of the union) to the rest of the company on a regular basis. Specifically, the network briefing is a monthly open dialog between the technical departments and the remaining company departments. The network briefing enables a large volume of information to be conveyed. The large volume of network information conveyed through the network briefings enables a uniform dissemination of information to take place.

It is recommended that the network briefings take on a format that is uniform and also timely for the general audience that will attend. The network briefings should take place on a monthly basis at a regular, predetermined time. The meeting should be scheduled to last no longer than two hours at most. The choice of using an overhead presentation or a more informal approach is more dependent upon your company's culture.

The format for the network briefing is as follows:

1. Introduction

   Introduce the various speakers and the agenda to be used for the meeting.

2. Engineering Activities

   This is where the last month's activities are talked about, usually a follow up to the previous month's comments. In addition, the current month's activities are discussed with expected outcomes. The general nature of the talk is high-level and not detail-oriented.

3. Construction

   This part of the discussion involves a representative from implementation and real estate, who talk about the current build program for the company. The topics to cover involve the recent completion of sites for the network. In addition to the sites built, a list of the sites currently under construction is also discussed.

4. Planned Network and Technical Projects

   The facilitator of the meeting goes over the projects and major events that are planned to take place between this meeting and the next with respect to the network.

5. Discussion

   This is the section in which members of the audience ask questions regarding any network issues.

6. Closing

The meeting is closed and the next meeting date and location are mentioned.

System Performance Troubleshooting

System performance and troubleshooting in an LMDS system involves various platforms and situations that span across multiple disciplines, requiring the proverbial team approach to resolving many of the issues that pass beyond the simple turn-the-power-on response. System performance and troubleshooting involves applying a set of critical techniques that have passed the test of time. The first technique to utilize is to identify your objective with the effort you are about to partake in and document it. The second technique is to isolate the item you are working on from the other variable parameters involved with the mission statement. The third technique is to identify what aspect of the system you are trying to work on: circuit-switching, packet-switching, Telco (CLEC), hub site, backbone, point to point radios, edge terminals, the RF environment, or the customers CPE. The fourth technique is to establish a battle plan: write down what you want to accomplish, how you will accomplish it, and what the expected results will be. The fifth technique is to communicate what you are about to do and why. The sixth technique is to conduct the work or troubleshooting that is identified in your objective, usually a test plan. The seventh technique is to conduct a post-analysis of your work and then issue a closing document, either supporting or refuting your initial conclusions and identifying the next actionable items.

Summarizing the system performance and troubleshooting methodology:

1. Identify objective

2. Remove variables

3. Isolate system components

4. Test plan

5. Communicate

6. Act

7. Conduct a post-analysis

This sequence can be used for any situation. Often, when working on a problem, the forest is lost in the trees and it is important to step back and re-evaluate the situation after a defined time period if the problem is not resolved. I have never encountered a situation that could not be resolved when these seven steps are applied to the problem or topic at hand.

References

Smith, Clint and Curt Gervelis. *Cellular System Design and Optimization*. New York, NY: McGraw-Hill, 1996.

McDysan, David E. and Darren L. Spohn. *ATM Theory and Application*, Signature ed. New York, NY: McGraw-Hill, 1998.

Spectrum
Allocation
Tables

Introduction

The spectrum allocations that have been allocated or are in the process of being allocated can be confusing to even experienced wireless professionals. Whenever a project begins or is about to begin, the one overriding questions that come up for any wireless system is the following: What is the spectrum that can be used and are there any restrictions?

The question of spectrum allocation and the associated tables that accompany it are provided in the next few sections for LMDS, FWPMP, WCS, ISM, MMDS/MDS/IFTS, GWCS, and PCS (U.S.) bands. What is interesting is that LMDS is offered in several countries, such as the 28GHz band, but the specific band allocations are not consistent. It is hoped that the information provided in this chapter will help shed some light on this important but complex subject.

LMDS (United States)

LMDS systems cover multiple frequency bands and, as discussed in previous chapters, they also include MMDS and FWPMP systems. The following sections show the spectrum breakout for several LMDS bands in the United States.

24GHz

The 24GHz band in the United States and Canada is allocated in 50MHz-paired blocks. Depending on the market involved, the aggregate spectrum for the license holder could range from 100Mhz to 400Mhz of spectrum.

A subchannel breakout is shown in Figure 11-1, in which the channels are allocated in 10Mhz-blocks that are paired.

28GHz

Table 11-1 shows the 28GHz spectrum-allocation scheme. There are a total of two licenses awarded for each BTA license area, one with the A block and the other the B block, for a total of 986 licenses in 493 basic trading areas.

The 28GHz spectrum allocation is shown in Figure 11-2.

Figure 11-1
24GHz spectrum allocation

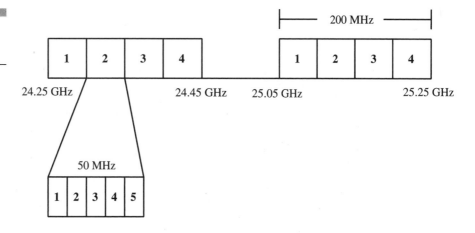

Table 11-1

28GHz

Block	Bandwidth	Frequencies
A	1150MHz	27,500–28,350MHz, 29,100–29,250MHz, and 31,075–31,225MHz bands
B	150MHz	31,000–31,075MHz and 31,225–31,300MHz bands

Figure 11-2
28GHz spectrum allocation

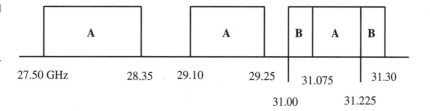

LMDS transceivers are prohibited from transmitting in the 29,100MHz–29,250MHz band because of concerns that theLMDS transceiver may interfere with NGSO-MSS feeder link space station receivers.

39GHz

The 39GHz band is located between 38.6GHz and 40GHz as shown in Figure 11-3. The channel blocks are all 50MHz in size, and there will be 14

Figure 11-3
LMDS 39GHz
spectrum

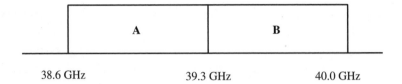

A	**B**	

38.6 GHz   39.3 GHz   40.0 GHz

Table 11-2

39GHz

Channel No.	Frequency Band Limits (MHz)	Channel No.	Frequency Band Limits (MHz)
1-A	38,600–38,650	1-B	39,300–39,350
2-A	38,650–38,700	2-B	39,350–39,400
3-A	38,700–38,750	3-B	39,400–39,450
4-A	38,750–38,800	4-B	39,450–39,500
5-A	38,800–38,850	5-B	39,500–39,550
6-A	38,850–38,900	6-B	39,550–39,600
7-A	38,900–38,950	7-B	39,600–39,650
8-A	38,950–39,000	8-B	39,650–39,700
9-A	39,000–39,050	9-B	39,700–39,750
10-A	39,050–39,100	10-B	39,750–39,800
11-A	39,100–39,150	11-B	39,800–39,850
12-A	39,150–39,200	12-B	39,850v39,900
13-A	39,200–39,250	13-B	39,900–39,950
14-A	39,250–39,300	14-B	39,950–40,000

paired blocks (100MHz of spectrum with a total of 175 licenses being awarded for each block). The channel spacing is such that the channel blocks have 700MHz spacing between the Tx and Rx channels.

The channel chart shown in Table 11-2, from the FCC Web site, helps clarify the channel blocks.

	Canadian Channel Band Designation	U.S. Band Designation
Table 11-3	B	3A, 3B
United States and	C	7A, 7B
Canadian 39GHz	D	8A, 8B
band	E	9A, 9B
	F	10A, 10B

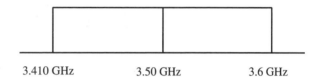

Figure 11-4
3.5GHz spectrum
allocation

3.410 GHz   3.50 GHz   3.6 GHz

In Canada, a similar channel plan and spectrum allocation is followed for the 39GHz band. The corresponding channel designations for the United States are shown next to the U.S. designations in Table 11-3.

FWPMP

Fixed Wireless Point to Multipoint (FWPMP) is the LMDS equivalent service outside of North America. The channel plan for FWPMP currently involves multiple bands, with more to come in the near future. The channel bandwidths are multiples of 1.75MHz, usually starting with 3.5MHz and 7MHz for usable channels.

3.5GHz

The 3.5GHz band has the distinct advantage over other LMDS and FWPMP frequencies for range. However, the range advantage is offset somewhat by the spectrum allocation. The 3.410 to 3.5GHz band is paired with the corresponding 3.5 to 3.6 GHz band. See Figure 11-4.

10GHz

The 10GHz band is another of the potential FWPMP bands to operate within as shown in Figure 11-5. The spectrum is paired and the Tx to Rx channel spacing is 350 MHz. See Table 11-4.

26GHz

The 26GHz band is one of the most interesting FWPMP bands, especially in Europe. The channel spacing between Tx and Rx is 1008MHz. The channels are usually assigned in increments of 7MHz. See Table 11-5.

Figure 11-5
10GHz spectrum allocation

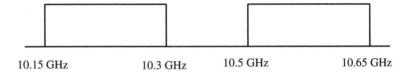

| 10.15 GHz | 10.3 GHz | 10.5 GHz | 10.65 GHz |

Table 11-4

10GHz

Channel Bandwidth (MHz)	Number of Usable Channels
3.5	42
7	20
14	10
28	5

Table 11-5

26GHz

Channel Bandwidth (MHz)	Number of Usable Channels
7	128
14	64
28	32
56	16
112	8

Figure 11-6
26GHz spectrum
allocation

24.5 GHz   25.445 GHz   25.557 GHz   26.5 GHz

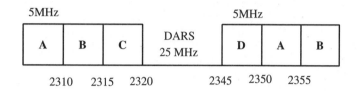

Figure 11-7
WCS spectrum
allocation

Not all channels are available for FWPMP, but the channel plan has been established. See Figure 11-6.

Wireless Communication Service (WCS)

Blocks A and B are assigned for 52 *major economic areas* (MEAs), with blocks C and D in 12 *regional economic area groups* (REAGs). The channel bandwidth for blocks A and B is 10MHz; for blocks C and D, it is 5MHz each. See Figure 11-7.

MDS/MDS/IFT

The MMDS channels are allocated in 6MHz channels. The channel allocations are shown in Table 11-6 for easy reference.

GWCS

General Wireless Communication Service (GWCS) has five 5MHz blocks assigned in each of the 175 economic areas defined as shown in Figure 11-8.

Table 11-6

MMDS/MDS/IFTS
Channels

Type	Channel Designation	Channel (MHz)
MDS	1	2150–2156
MDS	2	2156–2162
IFTS	A1	2500–2506
IFTS	B1	2506–2512
IFTS	A2	2512–2518
IFTS	B2	2518–2524
IFTS	A3	2524–2530
IFTS	B3	2530–2536
IFTS	A4	2536–2542
IFTS	B4	2542–2548
IFTS	C1	2548–2554
IFTS	D1	2554–2560
IFTS	C2	2560–2566
IFTS	D2	2566–2572
IFTS	C3	2572–2578
IFTS	D3	2578–2584
IFTS	C4	2584–2590
IFTS	D4	2590–2596
MMDS	E1	2596–2602
MMDS	F1	2602–2608
MMDS	E2	2608–2614
MMDS	F2	2614–2620
MMDS	E3	2620–2626
MMDS	F3	2626–2632
MMDS	E4	2632–2638
MMDS	F4	2638–2644
IFTS	G1	2644–2650
IFTS	H1	2650–2656
IFTS	G2	2656–2662
IFTS	H2	2662–2668
IFTS	G3	2668–2674
IFTS	H3	2674–2680
IFTS	G4	2680–2686

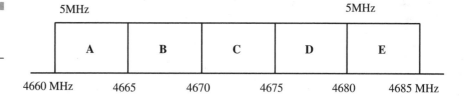

Figure 11-8
GWCS spectrum
(MHz)

Table 11-7

PCS band

PCS Block	Bandwidth (MHz)	Lower (MHz)	Upper (MHz)
A	30	1850–1865	1930–1945
D	10	1865–1870	1945–1950
B	30	1870–1885	1950–1965
E	10	1885–1890	1965–1970
F	10	1890–1895	1970–1975
C	30	1895–1910	1975–1990

PCS

The PCS spectrum allocations are mentioned here due the potential to offer broadband services with the spectrum allocated. The allocations are for a duplexed band that has a Tx-to-Rx channel separation of 80MHz if used for duplex communication.

The PCS blocks referenced are associated with the type of license holder that can exist in any market. The A, B, and C blocks have 30MHz of spectrum each, whereas the D, E, and F blocks each have 10MHz of spectrum. See Table 11-7.

UNII Band

Unlicensed National Information Infrastructure (UNII) is a unique band that is unlicensed and should be used with an LMDS license as a dual deployment strategy.

Figure 11-9
UNII band spectrum
allocation

5.25   5.35   5.725   5.825

Figure 11-10
ISM band spectrum
allocation

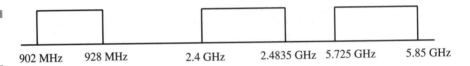

902 MHz   928 MHz   2.4 GHz   2.4835 GHz 5.725 GHz   5.85 GHz

The UNII band falls into the 5GHz to 5.825GHz band, 5.1GHz to 5.25GHz band, and 5.25GHz to 5.35GHz. The spectrum scheme is shown in Figure 11-9 for reference.

ISM Band

The *Instrumentation, Scientific and Medical* (ISM) band is located in many parts of the spectrum, but three most relevant to LMDS deployments are 902–928MHz, 2.4–2.4835GHz, and 5.725–5.85GHz, where the 2.4GHz part of the ISM band falls under shared microwave. See Figure 11-10.

Figure 11-11
700MHz band
plan [2]

700 MHz Band Plan

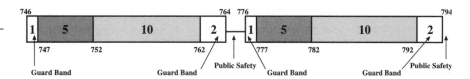

TV Channelization

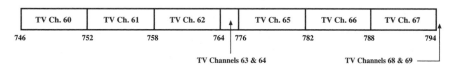

700MHz Band

The reallocation of the TV channels 60, 61, 61, 65, and 67 opens up a new element for LMDS operation in that LOS is not a predominant design criterion.

The spectrum is located in the 747–762MHz and 777–792MHz Bands. It is anticipated that the spectrum will be divided so that there will be six 20MHz licenses (10MHz paired) and an additional six having 10Mhz of spectrum each.

The current spectrum scheme is shown in Figure 11-11 for reference.

References

Smith, Clint. *Practical Cellular and PCS Design*. New York, NY: McGraw-Hill, 1997.

Federal Communications Commission: www.fee.gov

APPENDIX A

The proliferation of wireless platforms has also unleashed a much vaster array of three- and four-letter words within the industry. The rates at which these terms and definitions increase are faster then developments involving broadband access. Many times, the acronyms are vendor-specific, and other times they are industry-related or defined by the latest marketing push.

The list of acronyms contained here is by no means complete or all-encompassing. However, this is an attempt to list most of the relevant terms to facilitate daily operation within an LMDS operating company.

A

access network It is a portion of a public-switched network that connects access nodes to individual subscribers. The access network today is predominantly passive twisted-pair copper wiring.

access nodes They are points on the edge of any access network that concentrate individual access lines into a smaller number of feeder lines. Access nodes can also perform various forms of protocol conversion.

***Asymmetric Digital Subscriber Line* (ADSL)** ADSL is an xDSL technology in which modems are attached to twisted pair copper wiring that transmit from 1.5Mbps to 9Mbps downstream (to the subscriber) and from 16kbps to 800kbps upstream, depending on line distance between the subscriber and the wiring center.

***ATM Passive Optical Network* (APON)** It is a passive optical network running ATM.

asynchronous It is a form of communication in which there is no mandatory timing between two signals, hence the term asynchronous.

***Asynchronous Transfer Mode* (ATM)** ATM is an ultra-high-speed, cell-based data transmission protocol that is typically used as a bandwidth-efficient transport protocol. Several LMDS systems utilize ATM signaling as well as the core network for many PTTs.

B

backbone It is the primary trunk carrying voice or data traffic between switches.

bandwidth In simple terms, it is the size of the pipe through which data travels, or the amount of transport capacity a system or part of a system has. When discussing frequency, it is measured in hertz, or the number of cycles per second. When talking about data, the measure is bits per second (bps).

baud rate   The baud rate is the actual symbol frequency being used to transmit data.

bit   A bit is the smallest unit of information in a binary number system. Eight bits are needed to create one byte or character.

bits per second **(bps)**   It is the speed at which bits are transmitted across a data connection.

broadband   It is a term often used generically to describe systems with lots of bandwidth. Some prefer a more precise definition, such as systems with capacity of 1Mbps or greater.

BRI   BRI is a term used to describe ISDN signaling, in which a BRI consists of a 56kbps channel with a D-channel consisting of 6kbps of signaling. This is usually referred to as (1B+D); when two BRIs are used, the convention is to refer to it as (2B+D).

Broadband Integrated Digital Network **(B-ISDN)**   B-ISDN is a digital network with ATM-switching, operating at data rates in excess of 1.544 or 2.048Mbps. ATM enables the transport and switching of voice, data, image, and video over the same infrastructure.

bulletin board system **(BBS)**   BBS is the preferred method for many computer users to share and transfer information before the ascendance of the Internet. It is a centralized computer into which users typically dialed-in for exchange of files and messages.

byte   A byte is eight bits of information that represent one data character.

C

Carrier Access Code **(CAC)**   CAC is a dial sequence needed for a user to access switched services of an interexchange carrier.

Carrier Identification Code **(CIC)**   CIC is a unique four-digit number that identifies either the ILEC or CLEC.

Community Access Television **(CATV)**   CATV is also known as Cable TV.

central office **(CO)**   A CO is a secure building containing key equipment, such as switches, necessary for voice and data networks. CLECs have the statutory right to collocate and interconnect their equipment in an ILEC's central office.

Certificate of Public Convenience and Necessity **(CPCN)** CPCN is a document filed with the state public utility commission that certifies a company as a carrier.

channel A channel is a broad term referring to the pathway between two locations on a voice or data network. This term is often used to describe the radio assignments that exist in a wireless system (channel 1A).

channel bank It is the equipment used to combine or multiplex multiple channels either on a frequency-division or time-division basis.

Channel Service Unit/Data Service Unit **(CSU/DSU)** A CSU/DSU is the quipment on the customer's premises that handles digital transmissions.

circuit A circuit is simply a line that connects devices.

Competitive Local Exchange Carrier **(CLEC)** A CLEC is the wireline or wireless provider that can deliver alternative dial tones and other services using an incumbent PTT.

Customer Premises Equipment **(CPE)** The CPE is the equipment that is owned and operated by the customer. The term CPE is a customer's PBX, mobile phone, etc.

colocation Colocation is the placement of network equipment on the premises of another telecommunication service provider. It can be fixed network equipment or radio equipment.

co-provider Co-provider is another term for CLEC.

core network It is a combination of switching offices and transmission plant connecting switching offices together. In the United States, local exchange core networks are linked by several competing interexchange networks; in the rest of the world (now), the core network extends to national boundaries.

Carrier Serving Area **(CSA)** A CSA is an area served by an LEC, RBOC, or telco; often using *Digital Loop Carrier* (DLC) technology.

D

Digital Access Cross-connect **(DAC)** This equipment provides the ability to automatically interconnect (without manual wiring, by use of software-controlled hardware) individual channels of a T1/E1 on a one-by-one basis. This functionality allows for grooming and filling of the input T1/E1 spans to the DAC unit.

data protocols They are protocols or standards that have formal descriptions of rules and conventions that govern how devices on a network exchange information. Common protocols include TCP/IP, IPX, T1, E1, Frame Relay, ATM, etc.

Data Service Units **(DSUs)** This equipment amplifies an incoming signal from the PSTN for re-transmittal to the application equipment. In many cases this application equipment is a system switch.

Dynamic Host Configuration Protocol **(DHCP)** DHCP is a service supplied by a server that lets clients on a LAN request configuration information, such as IP host addresses.

dial-up It is the classical method of using the *Public Switched Telephone Network* (PSTN) to connect computers over modems to an ISP or LAN.

domain It is part of an Internet naming hierarchy. An Internet domain name consists of a sequence of names (labels) separated by periods (for example, ccseng.com).

Domain Name Service **(DNS)** A DNS is a TCP/IP protocol for discovering and maintaining network resource information distributed among different servers. It provides the ability to utilize a name for an IP address such as ccseng.com instead of 192.84.121.5.

download Download is the process of transferring a file from a server to a client.

Digital Subscriber Line Access Multiplexer **(DSLAM)** A DSLAM is a device that takes a number of ADSL subscriber lines and concentrates them to a single ATM line.

DS0 It is the portion of a T1/E1 that represents 64kbps worth of bandwidth.

Digital Subscriber Line **(DSL)** Usually referred to as xDSL, it can represent a vast variety of DSL protocols such as ADSL, HDSL, ISDL, VDSL, etc. DSL is a high-bandwidth, copper wire technology primarily for data.

Dual Tone Multi-Frequency **(DTMF)** It is a method of line address signal that uses two of seven in-band tones.

E

e-mail It means electronic messaging.

E1 An E1 is a 2.048Mbps channel that can be channelized or unchannelized, depending on the signaling to be transported. An E1 typically has 30 usable DS0s associated with it.

E & M signaling   It is a method of conveying supervisory and address signaling on a trunk by means of two signaling leads called the "E lead" and the "M lead." Signals are transmitted on the M lead and received on the E lead. There are several varieties of E & M signaling; they differ depending upon the application they are used within.

erlang   It is the unit of measure for telephone traffic used throughout the telecommunications industry. A single circuit that is occupied continuously over a one-hour period is equal to one erlang.

Ethernet   The Ethernet is the physical medium for transmitting *local area network* (LAN) traffic at speeds up to 10Mbps and 100Mbps.

F

facility   It is a medium of transmission that connects two fixed points in a network to provide communication between network elements. This medium can be a pair of copper wireline circuits, a fiber optics circuit, a microwave radio circuit, etc.

Far End CrossTalk **(FEXT)**   It is the interference occurring between two signals at the end of the lines remote from the telephone switch.

File Transfer Protocol **(FTP)**   FTP is a protocol that enables transfer files between computers. A common use of FTP on the Internet is to download software programs.

firmware   Firmware is a system software stored in a device's memory that controls the device.

Fiber To The Cabinet **(FTTCab)**   FTTCab is a network architecture in which an optical fiber connects the telephone switch to a street-side cabinet. The signal is converted to feed the subscriber over a twisted copper pair.

Fiber To The Home **(FTTH)**   It is a network in which an optical fiber runs from telephone switch to the subscriber's premises or home.

Fiber To the Curb **(FTTC)**   It is a network in which an optical fiber runs from telephone switch to a curbside distribution point close to the subscriber where it is converted to a copper pair.

G

gateway   A gateway is any system that transfers data between applications or networks that use different protocols.

glare condition   It is a condition that exists on a two-way trunk when it is seized (chosen by a switch for transmitting information) from both ends.

H

***High data Rate Digital Subscribe Line* (HDSL)**   HDSL is part of the xDSL family.

***Hybrid Fiber Coax* (HFC)**   It is a system in which fiber is run to a distribution point close to the end user or service facility and then the signal is converted to run to the subscriber's premises over coaxial cable.

home page   A home page is usually the starting page on a web site, such as www.ccseng.com.

host   A host is a single addressable device on a network. Computers, networked printers, and routers are referred to as hosts.

hub   A hub is a LAN device that connects computers, servers, and peripherals by repeating signals from one computer to the others on the LAN.

I

ISDL   ISDL uses ISDN transmission technology and is part of the xDSL family.

***Internet Service Provider* (ISP)**   An ISP is a service provider that connects users to the Internet.

***Incumbent Local Exchange Carriers* (ILECs)**   ILECs are the companies that were in business before the CLECs. The highest-visibility ILECs are the *regional Bell operating companies* (RBOCs).

***Integrated Switched Digital Network* (ISDN)**   An ISDN is a switched data network that handles both voice and data. Increasingly out of favor as faster technologies, such as DSL, become more widely deployed.

interconnection agreement   It is the broad agreement that determines how one wireline or wireless operator connects to another wireline or wireless operator.

Internet   It is the worldwide network of networks connected to each other by using the IP protocol suite.

intranet   It is an internal Internet designed to be a private network that is used within the confines of a company, university, or organization. What distinguishes an intranet from the freely accessible Internet is that intranets are private.

IP   It is the network-layer protocol for the Internet protocol.

J–K

kilobits per second (**Kbps**)   Kbps equals one thousand bits per second.

L

Local Area Network (**LAN**)   A LAN enables communication between computers and the sharing of local resources such as printers, CD readers, databases, and file servers.

Local Exchange Carrier (**LEC**)   A LEC is aso referred to as the local phone company.

Local Access and Transport Area (**LATA**)   A LATA is a geographical area—the boundaries of which are referred to as service areas.

local loop   It is the circuit between the end user and the central office. Sometimes referred to as "last mile" access.

M

megabits per second (**Mbps**)   Mbps equals millions of bits per second.

MF signaling   It is the high-speed (in-band) two of six analog tone trunk-signaling system that is used primarily for address signaling in the North American inter-toll network.

modem   It is a piece of data-communications equipment that converts a computer's digital signals to analog signals that can be transmitted over standard telephone lines.

Motion Picture Experts Group (**MPEG**)   There are a series of MPEG standards for compressed video transmission.

multiplexer (**MUX**)   A MUX is a device that combines two or more data streams into a single stream, providing 3/1 or 1/0 multiplexing. DACS and channel banks are also MUXs.

multimedia   It is anything using more than one medium: graphics, sound, animation, text, and/or video generated by a computer into one "presentation" on screen.

N

Network Address Translation (**NAT**)   It is the IP address from the private side of the network that is converted by a device for access to the public Internet.

Narrowband ISDN **(N-ISDN)** N-ISDN is the same as ISDN (1B+D).

Network Termination Equipment **(NTE)** NTE is equipment at the ends of the line from the wireless or wireline operator (that is, the smart jack or host terminal).

network Network is a generic term that refers to interconnected lines, switches, servers, software, and hardware that make up either a data or voice system.

number portability It is the capability for phone numbers issued by an ILEC to work for CLEC customers. It is a key competitive issue because customers are unlikely to switch to a CLEC if they must change phone numbers.

O

Optical Carrier 3 **(OC3)** OC3 is an optical fiber line carrying 155Mbps.

Optical Network Unit **(ONU)** ONU is a form of access node that converts optical signals to electrical signals.

P

Plain Old Telephone Service **(POTS)** POTS is the only name recognized around the world for basic analog telephone service.

Point of Presence **(POP)** POP is the location at which the telecommunications service provider has an access point where other service providers can possibly connect.

Point-to-Point Protocol **(PPP)** PPP provides router-to-router and host-to-network connections over both synchronous and asynchronous circuits.

Private Brach Exchange **(PBX)** PBX is the telephone-switching equipment located on the customer's premises that provide features and some call treatment.

Public Switched Telephone Network **(PSTN)** PSTN refers commonly to the RBOCs.

PTT PTT Refers to the incumbent state-controlled telephone company.

Public Utility Commission **(PUC)** It is a state commission that is sometimes called the Public Service Commission. It regulates all types of utilities, including the telecommunications industry.

Q–R

Rate Adaptive ADSL **(RADSL)**   It is a version of ADSL in which the ADSL modems adapt their operating speed to the fastest speed the line can handle.

Regional Bell Operating Company **(RBOC)**   An RBOC is one of the seven U.S. telephone companies that resulted from the break up of AT&T.

reciprocal compensation   It is payments that flow between the ILEC and the CLEC. The party that terminates the call receives the payment.

Routing Information Protocol **(RIP)**   RIP is a protocol used for the transmission of IP routing information.

routing table   It is a list contained in a router that helps the router determine the next router to forward packets to.

router   It is a device that distributes packet data traffic to the proper destination, based on the way it is programmed.

S

Symmetric Digital Subscriber Line **(SDSL)**   SDSL is another member of the xDSL family. As the name implies, the data rate it utilizes is symmetrical.

STS-1   STS-1 is the SONET basic transmission rate of 51.84Mbps.

Signaling System 7 **(SS7)**   SS7 is a protocol for addressing calls.

slamming   Slamming is a term used to describe the process of converting a customer from one carrier to another without the customer's permission.

switch   A switch is a device that connects callers to their destinations.

Synchronous Optical Network **(SONET)**   SONET is the technology at each end of an optical fiber system that connects, or interfaces, those fibers to the rest of a system.

serial port   It is any connector or port through which data flows to and from a serial device.

server   It is a device or system that has been specifically configured to provide a service, usually to a group of clients.

Simple Network Management Protocol **(SNMP)**   SNMP is a protocol used for communication between management consoles and network devices.

T

T1 T1 is the same as a DS1.

telco It is the generic name used to reference RBOCs, LECs or PTTs.

Telnet It is a virtual terminal protocol that allows users of one host to log into a remote host and interact as normal terminal users of that host.

Thicknet It is another name for 10Base-5 coaxial cable.

Thinnet It is another name for 10Base-2 coaxial cable.

trunk It is a facility that connects a subscriber unit to a network switch or a switch-to-switch connection, a switch-to-tandem connection, etc.

trunk group Trunk groups are groups in which trunks having similar characteristics and interconnecting points are arranged.

TCP/IP TCP/IP is a common name for the protocol developed by the U.S. Department of Defense in the 1970s to support construction of worldwide internetworks. As an open network standard, it defines how devices from different manufacturers communicate with each other over one or more interconnected networks.

U

Universal Service It is defiend as the responsibility of the wireline and some LMDS operators to provide service to virtually all customers in their geographic area, including those in low-profit areas such as rural areas.

V

***Very High Data Rate Digital Subscriber Line* (VDSL)** VDSL is part of the xDSL family, but operates at data rates from 12.9 to 52.8Mbps over copper wire.

Voice grade Voice grade is defined as channels that are suitable for good-quality voice transmissions. Such channels often also are used for fax and analog or digital data transmissions.

***Voice-over-DSL* (VoDSL)** It means to send voice transmissions over DSL networks.

***Voice-over IP* (VoIP)** It means to convert voice transmissions to IP packets of data and then transmit them over an IP network such as the Internet or a VPN.

***Virtual Private Network* (VPN)**   VPN describes the connection path from point A to point B within a communication network that provides a secure method of transmitting data and other information.

W–Z

***Wide Area Network* (WAN)**   WAN usually describes a data network that involves connecting several campus locations together, either within the same city or between cities.

INDEX

T

ABOUT THE AUTHOR

Clint Smith, P.E., is vice-president of Operations for 02 Wireless Solutions and was the former vice-president of Technical Services for Communication Consulting Services and formally Director of Engineering at NYNEX Mobile Services. He is also the author of *Cellular System Design and Optimization*, *Practical Cellular and PCS Design*, and *Wireless Telecom FAQ's*.